The Rice-Wheat Cropping System of South Asia: Trends, Constraints, Productivity and Policy

The Rice-Wheat Cropping System of South Asia: Trends, Constraints, Productivity and Policy has been co-published simultaneously as *Journal of Crop Production*, Volume 3, Number 2 (#6) 2001.

The *Journal of Crop Production* Monographic "Separates"

Below is a list of "separates," which in serials librarianship means a special issue simultaneously published as a special journal issue or double-issue *and* as a "separate" hardbound monograph. (This is a format which we also call a "DocuSerial.")

"Separates" are published because specialized libraries or professionals may wish to purchase a specific thematic issue by itself in a format which can be separately cataloged and shelved, as opposed to purchasing the journal on an on-going basis. Faculty members may also more easily consider a "separate" for classroom adoption.

"Separates" are carefully classified separately with the major book jobbers so that the journal tie-in can be noted on new book order slips to avoid duplicate purchasing.

You may wish to visit Haworth's website at . . .

http://www.HaworthPress.com

. . . to search our online catalog for complete tables of contents of these separates and related publications.

You may also call 1-800-HAWORTH (outside US/Canada: 607-722-5857), or Fax 1-800-895-0582 (outside US/Canada: 607-771-0012), or e-mail at:

getinfo@haworthpressinc.com

The Rice-Wheat Cropping System of South Asia: Trends, Constraints, Productivity and Policy, edited by Palit K. Kataki, PhD (Vol. 3, No. 2 #6, 2001). *This book critically analyzes and discusses available options for all aspects of the rice-wheat cropping system of South Asia, addressing the question, "Are the sustainability and productivity of this system in a state of decline/stagnation?" This volume compiles information gathered from research institutions, government organizations, and farmer surveys to analyze the impact of this regional system.*

Nature Farming and Microbial Applications, edited by Hui-lian Xu, PhD, James F. Parr, PhD, and Hiroshi Umemura, PhD (Vol. 3, No. 1 #5, 2000). *"Of great interest to agriculture specialists, plant physiologists, microbiologists, and entomologists as well as soil scientists and evnironmentalists. . . very original and innovative data on organic farming." (Dr. André Gosselin, Professor, Department of Phytology, Center for Research in Horticulture, Université Laval, Quebec, Canada)*

Water Use in Crop Production, edited by M.B. Kirkham, BA, MS, PhD (Vol. 2, No. 2 #4, 1999). *Provides scientists and graduate students with an understanding of the advancements in the understanding of water use in crop production around the world. You will discover that by utilizing good management, such as avoiding excessive deep percolation or reducing runoff by increased infiltration, that even under dryland or irrigated conditions you can achieve improved use of water for greater crop production. Through this informative book, you will discover how to make the most efficient use of water for crops to help feed the earth's expanding population.*

Expanding the Context of Weed Management, edited by Douglas D. Buhler, PhD (Vol. 2, No. 1 #3, 1999). *Presents innovative approaches to weeds and weed management.*

Nutrient Use in Crop Production, edited by Zdenko Rengel, PhD (Vol. 1, No. 2 #2, 1998). *"Raises immensely important issues and makes sensible suggestions about where research and agricultural extension work needs to be focused." (Professor David Clarkson, Department of Agricultural Sciences, AFRC Institute Arable Crops Research, University of Bristol, United Kingdom)*

Crop Sciences: Recent Advances, Amarjit S. Basra, PhD (Vol. 1, No. 1 #1, 1997). *Presents relevant research findings and practical guidance to help improve crop yield and stability, product quality, and environmental sustainability.*

The Rice-Wheat Cropping System of South Asia: Trends, Constraints, Productivity and Policy

Palit K. Kataki, PhD
Editor

The Rice-Wheat Cropping System of South Asia: Trends, Constraints, Productivity and Policy has been co-published simultaneously as *Journal of Crop Production*, Volume 3, Number 2 (#6) 2001.

CRC Press
Taylor & Francis Group
Boca Raton   London   New York

CRC Press is an imprint of the
Taylor & Francis Group, an informa business

Reprinted 2010 by CRC Press

CRC Press
6000 Broken Sound Parkway, NW
Suite 300, Boca Raton, FL 33487
270 Madison Avenue
New York, NY 10016
2 Park Square, Milton Park
Abingdon, Oxon OX14 4RN, UK

Published by

Food Products Press®, 10 Alice Street, Binghamton, NY 13904-1580 USA

Food Products Press® is an imprint of The Haworth Press, Inc., 10 Alice Street, Binghamton, NY 13904-1580 USA.

The Rice-Wheat Cropping System of South Asia: Trends, Constraints, Productivity and Policy has been co-published simultaneously as *Journal of Crop Production*, Volume 3, Number 2 (#6) 2001.

The development, preparation, and publication of this work has been undertaken with great care. However, the publisher, employees, editors, and agents of The Haworth Press and all imprints of The Haworth Press, Inc., including The Haworth Medical Press® and Pharmaceutical Products Press®, are not responsible for any errors contained herein or for consequences that may ensue from use of materials or information contained in this work. Opinions expressed by the author(s) are not necessarily those of The Haworth Press, Inc.

Cover design by Thomas J. Mayshock Jr.

Library of Congress Cataloging-in-Publication Data

The rice-wheat cropping system of South Asia: trends, constraints, productivity and policy/Palit K. Kataki, editor.
    p. c.m.
    "Co-published simultaneously as Journal of crop production, volume 3, number 2 (#6) 2001."
    Includes bibliographical references and index.
    ISBN 1-56022-084-8 (hardcover: alk. paper)–ISBN 1-56022-085-6 (pbk.: alk. paper)
    1. Rice–South Asia. 2. Wheat–South Asia. 3. Cropping systems–South Asia. I. Journal of crop production.
SB191.R5 R5355 2001
338.1'62'0954–dc21
                                                 00-068194

The Rice-Wheat Cropping System of South Asia: Trends, Constraints, Productivity and Policy

CONTENTS

Preface   xi

Foreword   xvii
    Timothy G. Reeves

The Rice-Wheat Cropping System of South Asia: Trends,
    Constraints and Productivity–A Prologue   1
    P. K. Kataki
    P. Hobbs
    B. Adhikary

Long-Term Yield Trends in the Rice-Wheat Cropping System:
    Results from Experiments and Northwest India   27
    J. M. Duxbury

An Agroclimatological Characterization of the Indo-Gangetic
    Plains   53
    J. W. White
    A. Rodriguez-Aguilar

Legumes and Diversification of the Rice-Wheat Cropping
    System   67
    J. G. Lauren
    R. Shrestha
    M. A. Sattar
    R. L. Yadav

Policy Re-Directions for Sustainable Resource Use:
    The Rice-Wheat Cropping System of the Indo-Gangetic
    Plains   103
        P. L. Pingali
        M. Shah

Synthesis of Systems Diagnosis: "Is the Sustainability
    of the Rice-Wheat Cropping System Threatened?"–
    An Epilogue   119
        L. Harrington

Index   133

ABOUT THE EDITOR

Palit K. Kataki, PhD, received his doctorate at Cornell University. He has a joint appointment as a Cornell University Coordinator and a CIM-MYT (Centro Internacional de Mejoramiento de Miaz y Trigo) Adjunct Scientist. He has been involved in the planning, implementation, and coordination of research on rice and wheat in South Asia (Bangladesh, India, Nepal, and Pakistan) for several years. Dr. Kataki's research emphasizes tillage and crop establishment, micronutrients, field crop sterility, the establishment of legumes in the rice-wheat cropping system, and seed physiology.

Dr. Kataki has served as Assistant Manager of a tea estate (Tata Tea, Ltd.), Research Associate at ICAR (Indian Council of Agricultural Research), and Assistant Professor at Assam Agricultural University. He has been the recipient of several awards and scholarships, and has authored or co-authored several scientific papers.

Preface

Only a few cropping systems satisfy the calorie needs of over a billion people within a sub-continent, the "Rice-Wheat Cropping System" practiced in the South Asian Indo-Gangetic Plains (IGP) region being one of them. The pros and cons of the past three decades of the largely successful "Green Revolution" era of the IGP are still being debated at various levels of society, from the philosophers to the active field agricultural scientist. Meanwhile, almost as if silently, new sets of problems and concerns have crept into the adoption and spread of this cropping system in the IGP, for both the rice and the wheat crop. These concerns include the need for changes in crop establishment techniques and for crop diversification, declining soil fertility, changes in the pest scenario, a host of issues relating to water management, and a need for policy redirection to sustain productivity growth in the IGP.

At the core, is a search for answers to the questions: "Is the productivity (and sustainability) of the rice-wheat cropping system stagnating OR declining?" AND "Do we have any evidence for it?" for the vast IGP region which encompasses four countries (Bangladesh, India, Nepal and Pakistan) across different agro-ecological zones. Therefore, the theme of this publication on the Rice-Wheat Cropping System of South Asia is centered on these questions. Answering on the affirmative for both these questions would mean understanding the constraints to system productivity and then seeking solutions. The topics in this publication therefore discusses the trends of rice-wheat over time, analyzes rice-wheat productivity changes, followed by specific issues on rice and wheat crop production, which are important in general for the entire IGP region. Topics important in certain areas of the IGP but couldn't be discussed in this special issue are–arsenic contamination

[Haworth co-indexing entry note]: "Preface." Kataki, Palit K. Co-published simultaneously in *Journal of Crop Production* (Food Products Press, an imprint of The Haworth Press, Inc.) Vol. 3, No. 2(#6), 2001, pp. xv-xix; and: *The Rice-Wheat Cropping System of South Asia. Trends, Constraints, Productivity and Policy* (ed: Palit K. Kataki) Food Products Press, an imprint of The Haworth Press, Inc., 2001, pp. xi-xv. Single or multiple copies of this article are available for a fee from The Haworth Document Delivery Service [1-800-342-9678, 9:00 a.m. - 5:00 p.m. (EST). E-mail address: getinfo@haworthpressinc.com].

of ground water in the eastern IGP, boron toxicity especially to animals in certain districts of Punjab, sulfur deficiency in soils, and weed management. A common denominator for all the issues discussed here is that of varieties, in other words, crop improvement strategies to breed for varieties will be issue and region specific, once the issues (problems) are clearly understood.

The rice-wheat cropping system is here to stay but there are signs of its productivity growth stagnating or declining. The total production of rice and wheat in the IGP has been increasing annually (chapter by P. K. Kataki, P. Hobbs and B. Adhikary). Under this scenario, it is often difficult to provide hard evidence of stagnating or declining productivity trends given the size (acreage) of the IGP region and the agro-climatic diversity within. The three data sources that provide evidence of productivity changes are from the farmer's field and benchmark surveys, the on-going location specific long-term monitoring (8 to 9 years) surveys, and research station managed long-term soil fertility experiments. The article by P. K. Kataki, P. Hobbs and B. Adhikary also compiles results and discusses the farmer's field and benchmark surveys and Total (TPF) and Partial Factor Productivity (PFP) analysis done for the region. Evidence till date indicate declining productivity. The chapter on the analyses of long-term soil fertility experiment station yields trends by J. M. Duxbury further supports this argument. Analysis of results from the long-term monitoring survey is not complete and is therefore not being discussed here. Focus has shifted from a "Commodity" type research program to diagnosing and understanding the "System" requirements (article by L. Harrington). This has encouraged greater inter-disciplinary research in the region and is evident from the topics covered in this special issue.

Agroclimatic characterization of the IGP is important for the diagnosis and analyses of "System" requirement (chapter by J. W. White and A. Rodriguez-Aguilar). Options for tillage and crop establishment techniques of rice and wheat have been experimented with and field-tested and are at various stages of adoption and spread in the farming community (chapter by P. Hobbs [volume 2 of the series]). These tillage options include region specific solutions based on the land holdings, level of mechanization, soil and climatic conditions, and socio-economic constraints, which varies widely within the IGP. The rice-wheat system has literally reduced the desire to maintain crop diversification within the IGP and has pushed the cultivation of beneficial legumes to more marginal areas in the region (chapter by J. G. Lauren, R. Shrestha, M. A. Sattar and R. L. Yadav). Legumes have been an integral part of South Asian diet and culture. This chapter discusses the options available to reintroduce legumes into the IGP to more than its present level of cultivation and to have a greater impact on the soil "health" and on human nutritional benefits.

Soils of the IGP are inherently low in soil fertility, and the further decline in soil fertility in the region is a regional concern. The primary reason for this decline is due to the present soil management practices, or rather its mismanagement in general. Except for the introduction of high yielding, dwarf, fertilizer responsive and disease resistant rice and wheat varieties and increased availability of irrigation water, the "culture" of growing rice and wheat in the region has in general, not adjusted with the demands of the new "commodity" in the last three decades. Higher the crop yield, greater is the nutrient uptake by plants for meeting the "sink" needs, and the requirement for available soil nutrients to sustain this yield jump increases. Higher crop yields (productivity, and therefore the total quantity or production) is absolutely essential to meet the calorie requirement of an ever-increasing population of the region and therefore adjusting crop management practices is important "in making both ends meet." Yadvinder-Singh and Bijay-Singh (volume 2 of the series) discuss the management of the primary nutrient requirement for rice and wheat in the IGP. Nitrogen is ubiquitously deficient and deficiencies of P and K are increasing in the region. Fertilizer policy has encouraged imbalance use of N:P:K in the region. Use of organic sources of nutrients has decreased in the IGP. Intensive cropping also increases micronutrient deficiencies. V. K. Nayyar, C. L. Arora and P. K. Kataki (volume 2 of the series) discuss the changes in the soil micronutrient status in the IGP, and its management in relation to rice and wheat. In the eastern IGP, soil Boron deficiency is widespread, reducing crop yields (chapter by P. K. Kataki, S. P. Srivastava, M. Saiffuzaman and H. K. Upreti [volume 2 of the series]).

The spectrum of pests on rice and wheat has changed since the adoption and spread of the "Green Revolution" in the IGP (article by M. Sehgal, M. D. Jeswani and N. Kalra [volume 2 of the series]). Pest attack can be visual (if above ground) or hidden (e.g., damage caused by nematodes, article by S. B. Sharma [volume 2 of the series]). The integrated management (and constraints to its adoption and spread) of pests are discussed in these chapters in relation to the IGP.

Inadequate water management practices in the rice-wheat cropping system of the IGP are a widely acknowledged constraint, but the desire to overcome these problems at various levels have fallen short of expectations. Appropriate water management has to be dealt with starting at the macro level through active policy changes, to the micro or field level in terms of crop water needs and its management. Water policy and use is too big a subject to cover all aspects in this publication. Therefore a case study of water management from the Sindh-zone of Pakistan by M. Aslam and S. A. Prathapar (volume 2 of the series), and use and management of poor quality water in the high yielding western IGP region (by P. S. Minhas and M. S. Bajwa [volume 2 of the series]) is being discussed here. Finally, policy framing and its adjustment (re-directions) to overcome all the constraints mentioned here, are important for sustainable resource use in the IGP in relation to the rice-wheat cropping system (chapter by P. L. Pingali and M. Shah).

All of the above topics are being published in two volumes. Volume 1 of the publication discusses 6 topics on the status of rice-wheat in the IGP and covers the following: Trends of the Rice-Wheat Cropping System, Long-Term Soil Fertility Experiments, Agro-Climatic Characterization of the IGP, Legumes and Crop Diversification, Policy Re-Directions for Sustainable Resource Use, and the Synthesis of Systems Diagnosis. Volume 2 of this series discusses the major constraints of the rice-wheat cropping system and includes: Tillage and Crop Establishment, Efficient Management of Primary Nutrients and Micronutrients, Sterility in Wheat and Field Crop Response to Applied Boron, Pest Management, Plant Parasitic Nematodes, Water Management in Sindh of Pakistan, and Use and Management of Poor Quality Waters.

The volume of literature cited in preparation for all the chapters in this publication is a testament to the work done over many years by the agricultural scientists within the network of National Agricultural Institutes, Universities, and International Agricultural Centers in the IGP region. This is specifically true for issues related to rice and wheat production within the last three decades. As a result, these Institutes were able to form the Rice-Wheat Consortium (see Foreword by Professor Timothy Reeves, Director General, CIMMYT) to facilitate greater co-operation, exchange of ideas, research results and research cooperation across the vast IGP region. The volume of work done is immense and this publication on rice-wheat attempts to comprehensively discuss the major problems of this important cropping system and strategies to overcome these problems, the target audience being researchers, students, extension personnel and policy makers.

For providing logistical and financial support for work on rice and wheat in the years of my involvement in the region, I would like to acknowledge the following: National Agricultural Institutes of Bangladesh, India, Nepal and Pakistan, the innumerable agricultural scientists working in these Institutes, International organizations (CIMMYT, ICRISAT, IRRI), Cornell University (especially the Department of Crops and Soils, and CIIFAD), US-AID SM-CRSP program, and the Rice-Wheat Consortium for the Indo-Gangetic Plains. Finally to Kalpita, Gemma, Ronnie and Rubin for their love and patience.

Palit K. Kataki

Foreword

The Rice-Wheat Consortium for the Indo-Gangetic Plains (RWC) was founded in 1993 to foster sustainable productivity in rice-wheat systems. The impetus for creating the RWC was a concern over declining productivity in rice-wheat systems, in which rice and wheat are grown in sequence on the same plot of land. This system meets the food needs of about one billion people: faltering productivity will exact a heavy toll. The rice-wheat sequence occupies some 12 million hectares in Bangladesh, India, Nepal, and Pakistan and accounts for about 90% of food production over this area. Future food security for the region's expanding population is threatened by a range of difficulties, including slower growth in rice and wheat yields, few options for expanding cropped area, and alarming signs of natural resource depletion and degradation.

The RWC has assembled scientific and technical experts from the national research systems of Bangladesh, India, Nepal, and Pakistan, international research centers, and advanced research institutes. The mission they share is to bring a multidisciplinary, holistic perspective to bear on the problems limiting sustainability in rice-wheat cropping systems. The RWC was established in 1994 as an Ecoregional Initiative of the Consultative Group on International Agricultural Research (CGIAR), and in 1998, the International Maize and Wheat Improvement Center (CIMMYT) was assigned convening and leadership responsibilities.

The priorities and research themes of the RWC, which are explored in detail in this volume, reflect the broad systems perspective of the research team:

[Haworth co-indexing entry note]: "Foreword." Reeves, Timothy G. Co-published simultaneously in *Journal of Crop Production* (Food Products Press, an imprint of The Haworth Press, Inc.) Vol. 3, No. 2(#6), 2001, pp. xxi-xxv; and: *The Rice-Wheat Cropping System of South Asia: Trends, Constraints, Productivity and Policy* (ed: Palit K. Kataki) Food Products Press, an imprint of The Haworth Press, Inc., 2001, pp. xvii-xxi. Single or multiple copies of this article are available for a fee from The Haworth Document Delivery Service [1-800-342-9678, 9:00 a.m. - 5:00 p.m. (EST). E-mail address: getinfo@haworthpressinc.com].

- long-term experiments to understand why productivity is declining in rice-wheat systems;
- alternative tillage and crop establishment practices (zero tillage, surface seeding of wheat, direct seeding of rice, bed planting, small-scale mechanization) to increase productivity, foster diversification in farming systems, and reduce production costs;
- integrated nutrient management practices (balanced nutrient use, use of chlorophyll meters, studies of micronutrients, new rotations, crop residue management) to obtain higher yields with less chemical fertilizer and rebuild the soil resource base;
- practices to stabilize yields in cropping systems and improve system ecology (alternative crops and rotations, integrated pest management, studies of long-term effects of new tillage practices) to foster long-term productivity, profitability, and diversity in cropping systems;
- integrated water management (practices to reduce salinity and sodicity, groundwater depletion, waterlogging) from the field level to the water basin level, to address projected water shortages and water management problems in coming decades in South Asia;
- the development of rice and wheat varieties to perform well in conservation tillage and direct sowing systems;
- attention to the socioeconomic and policy issues that affect the profitability of new practices, the adoption of new practices, and the rice and wheat economy of the region;
- data and geographic information systems (GIS) to extend the reach of science over years and across sites;
- networking across borders to unite previously isolated researchers and institutions with similar goals; and
- human resource development enabling researchers and farmers to develop, experiment with, and apply new technologies.

Research priorities are constantly evaluated and revised by the members of the RWC, in recognition of the fact that, to be sustainable, farming systems must be economically viable, environmentally sound, socially acceptable, and politically supportable.

Sustainable farming systems must be *economically viable* at the farm and national levels. Poor farmers cannot invest in systems that will not produce reasonable yields and (even better) cash income, now and in the future. At the national level, agriculture must also earn its keep as a significant contributor to GDP and export earnings. The reality in most countries, including those participating in the RWC, is that economic well being and development are almost invariably based on productive and profitable agriculture, the "engine room" of subsequent industrialization.

Sustainable farming systems must be *environmentally sound* as well. Economic success in agriculture cannot come at the expense of our soils, air, water, landscapes, and indigenous flora and fauna.

Sustainable farming systems must be *socially acceptable*. They must be appropriate to the people who, relying on their own meager resources, are responsible for implementing and managing them. The need for socially acceptable systems implies the need for a better understanding of farmer and community needs and values, as well as better targeting of technology to meet local conditions.

Finally, sustainable farming systems must be *politically support-able*. Political support depends largely on successfully meeting the first three requirements of sustainability. If economic growth is catalyzed by agriculture within an environmentally sound, socially acceptable framework, politicians will continue to view agriculture as justifying support.

All four components combine to form the whole: sustainable agriculture. If one is neglected, it can seriously reduce the rate and extent of progress towards sustainability and food security.

Another important contribution of the RWC to research in South Asia is the recognition that to attain sustainable food securities we must change the way we plan, conduct, and communicate about research. We must blend very specialized research disciplines in teams of scientists seeking appropriate outcomes that have an immediate impact in farmers' fields. The RWC, like CIMMYT, has adopted an integrative research paradigm that brings together the best genotypes (G), in the right environments (E), under appropriate crop management (M), generating appropriate outcomes for people (P). All of us who are dedicated to fostering sustainable agriculture in developing countries recognize the interdependence of these factors. Most organizations alone cannot contribute fully to each aspect of $G \times E \times M \times P$. Partnerships and consortia that assemble the best possible teams to execute the $G \times E \times M \times P$ approach, such as the RWC, will

underpin the timely and successful achievement of sustainable farming systems and future food security.

The chapters in this volume constitute a valuable overview of our
present knowledge of sustainability problems in the rice-wheat systems of South Asia. They present the results of long-term experiments
designed to gauge the nature and extent of sustainability problems.
They also describe the practices and technologies that have been developed, tested, and in some cases adopted to overcome constraints to
sustainability. Readers should note, however, that this volume is not
simply an account of the progress that the RWC has achieved to date,
as important as that is. This book is also a guide to what may be
achieved by using a research approach in which highly focused, multidisciplinary teams of researchers work closely with farmers. Two
examples of the impacts of this research approach will suffice to
illustrate this point; readers will find many others in the chapters that
follow.

FIRST

Two alternative tillage methods promoted by the RWC–direct drilling and surface seeding–allow farmers to prepare soils and sow wheat
in a single tractor operation after the rice harvest. These practices
enable farmers to reduce production costs by US$ 65/ha. It reduces
fuel use by 75% or more, provides better yields, cuts herbicide use in
half, and requires 10% less water. In 2000, farmers used direct drilling
with locally manufactured drills to plant 8,000 hectares in Haryana,
India, and 5,000 hectares in the Punjab of Pakistan. The area of adoption has increased ten-fold each year for several years. The main
constraint on more rapid expansion has been a lack of good quality,
fairly priced seed drills, but now small private shops are producing
more drills in response to the rising demand.

SECOND

The potential environmental impacts of the practices promoted by
the RWC are impressive. By changing to a zero-till system on one
hectare of land, a farmer can save 98 liters of diesel and approximately

1 million liters of irrigation water. Using a conversion factor of 2.6 kilograms of carbon dioxide per liter of diesel burned, this represents about a reduction in carbon dioxide emission of a quarter ton. Carbon dioxide is a principal contributor to global warming. These benefits increase dramatically if extended across even a portion of the rice-wheat region's 12 million hectares. Adoption of zero tillage on 5 million hectares would represent a saving of *5 billion cubic meters of water* each year. That would fill a lake 10 kilometers long, 5 kilometers wide, and 100 meters deep. In addition, annual diesel fuel savings would come to *0.5 billion liters*–equivalent to a *reduction of nearly 1.3 million tons in CO_2 emissions* each year.

These technologies give an idea of what can be done to achieve more sustainable agriculture in major production systems that are crucial to the welfare of large numbers of people. They also give an idea of the vast challenge taken on by the RWC. The RWC's work is certainly not finished, and readers will appreciate that the chapters in this volume describe a "work in progress" rather than a job completed. The pages that follow reflect the importance of the RWC's work, the dedication of the researchers in the Consortium, and the tremendous human impact of their work. Those of us from CIMMYT who have been privileged to share in this work are proud of the achievements to date. The contents of this volume serve as a testimony to the enduring value of the research and to the great urgency and importance of future support for this group of dedicated researchers.

Professor Timothy G. Reeves
Director General
International Maize and Wheat Improvement Center (CIMMYT)

The Rice-Wheat Cropping System of South Asia: Trends, Constraints and Productivity– A Prologue

P. K. Kataki
P. Hobbs
B. Adhikary

SUMMARY. The rice-wheat cropping pattern of South Asia has seen a phenomenal growth in area, production and yield in the last two decades. This cropping system encompasses the four countries of Bangladesh, India, Nepal and Pakistan along the Indo-Gangetic Plains (IGP) and into the mid-hills of the Himalayas. Traditionally, wheat was grown mostly in the northwestern belt and rice in the eastern belt of the IGP. With the introduction of improved high yielding, input responsive, short duration rice and wheat varieties, the rice-wheat pattern became feasible and saw both crops grown in the same year. In this pattern, rice is grown during the summer months followed by wheat in the winter months. It is now found as a major system throughout the IGP. However, there are signs that the productivity of this cropping system is either stagnating or declining. Several site-specific benchmark and diagnostic farm level surveys have been conducted in the region to understand the

---

P. K. Kataki is Cornell University Coordinator, and CIMMYT Adjunct Scientist, Nepal; P. Hobbs is Regional Representative and Agronomist for South Asia, CIMMYT, Nepal; and B. Adhikary is Biometrician, CIMMYT, Nepal.

Address correspondence to: P. K. Kataki, CIMMYT South Asia Regional Office, Lazimpat, P.O. Box 5186, Kathmandu, Nepal (E-mail: pkataki@vsnl.com).

[Haworth co-indexing entry note]: "The Rice-Wheat Cropping System of South Asia: Trends, Constraints and Productivity–A Prologue." Kataki, P. K., P. Hobbs, and B. Adhikary. Co-published simultaneously in *Journal of Crop Production* (Food Products Press, an imprint of The Haworth Press, Inc.) Vol. 3, No. 2(#6), 2001, pp. 1-26; and: *The Rice-Wheat Cropping System of South Asia: Trends, Constraints, Productivity and Policy* (ed: Palit K. Kataki) Food Products Press, an imprint of The Haworth Press, Inc., 2001, pp. 1-26. Single or multiple copies of this article are available for a fee from The Haworth Document Delivery Service [1-800-342-9678, 9:00 a.m. - 5:00 p.m. (EST). E-mail address: getinfo@ haworthpressinc.com].

constraints of this cropping system and ways to improve its productivity and sustainability. This article analyzes the trend in the area, production and yields of rice and wheat as a whole and assumes these trends are similar in rice-wheat areas. It also summarizes the results of the farm level surveys and productivity analyses done in the IGP on rice and wheat. *[Article copies available for a fee from The Haworth Document Delivery Service: 1-800-342-9678. E-mail address: <getinfo@haworthpressinc. com> Website: <http://www.HaworthPress.com> © 2001 by The Haworth Press, Inc. All rights reserved.]*

KEYWORDS. Bangladesh, cropping pattern, cropping system, Himalayas, India, Indo-Gangetic Plains, Nepal, Pakistan, productivity, rice, South Asia, wheat

INTRODUCTION

The "rice-wheat cropping system" (RWCS) region of South Asia spans the fertile Indo-Gangetic Plains (IGP) of four countries, from western Pakistan, northern, northwestern and eastern India, the Terai plains and portions of the Himalayan mid-hills of Nepal, to the western and northwestern region of Bangladesh (Figure 1). Traditionally, eastern India, Nepal and Bangladesh were rice-growing areas with very little wheat being cultivated, while in northwestern India and Pakistan, wheat was the major cereal crop with very little or no rice being grown in the early 1960's. The traditional varieties of rice and wheat were of longer duration; thereby requiring more time to reach physiological maturity for harvest, and were not responsive to intensive inputs and management. The green revolution introduced shorter duration, dwarf, largely photo-insensitive, high yielding, and fertilizer responsive varieties of rice and wheat during the 1960's and 1970's. Due to these varietal changes, it was possible to grow rice during the monsoon months following wheat in the otherwise non-rice growing northwestern region of these fertile plains. Simultaneously, it was also possible to plant wheat in November/December after rice harvest in the eastern region of the IGP. During the last two decades, the spread of this cropping system in this fertile region earned it the terminology "The Rice-Wheat Cropping System" region.

Over twenty cropping patterns are practiced in South Asia, but the rice-wheat cropping sequence dominates in area, production and productivity (Yadav, Prasad and Singh, 1998). Besides appropriate vari-

FIGURE 1. Schematic map of the Indo-Gangetic Plains (IGP) showing the rice-wheat growing regions (shaded areas) of Bangladesh, India, Nepal and Pakistan.

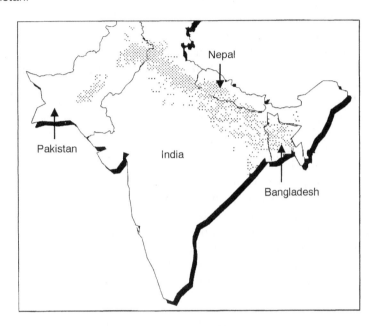

etal development, a key to the success of the rice-wheat rotation has been the development of irrigation facilities in the hotter and drier western region of the IGP. Regions with assured irrigation has favored the rice-wheat cropping system, while the rainfed and dryland regions support a larger variety of cropping patterns in addition to rice-wheat, due to the greater risks involved (Yadav, Gangwar and Prasad, 1998; Yadav, Prasad, Gangwar, 1998).

This article analyzes the trend in the area, production, and productivity of rice and wheat in the region, followed by a summary of the site-specific farm level surveys conducted in the region that identified the constraints in this cropping system. Finally, productivity analysis done on rice-wheat is discussed in the last section of this article. The data for the analysis of the acreage, production, and yield trends for this paper has been taken from country and international publications (multiple years) including the Fertilizer Association of India, Ministry of Agriculture, Nepal, Ministry of Agriculture, Pakistan, Bangladesh Bureau of Statistics, International Rice Research Institute, and FAO.

Statistical Issues

Unfortunately, production and yield statistics are not collected on a cropping system basis, but rather for individual crops of rice and wheat, and so specific data solely for the rice-wheat cropping system is not available. This would require aggregating District level data and making assumptions for each District on what percentage of each crop is in the rice-wheat system. That was not feasible and so the following data is based on statistics for the entire region and includes data from all cropping systems. It is estimated that there are 12 million hectares of rice-wheat in the region, 9 million in India, 1.5 million in Pakistan and the rest in Nepal and Bangladesh. In Nepal and Bangladesh 80-85% of the wheat follows rice, so wheat statistics should follow total statistics. However, only 30% and 5% of the rice in Nepal and Bangladesh is in the rice-wheat system. In India, only 21% of the rice area and 35% of wheat is rice-wheat, so some caution should be given to interpreting the following trends. In Pakistan, 18% and 70% of the wheat and rice, respectively, are in the rice-wheat areas. The assumption is made that growth rates in area, production and yield for those in rice-wheat systems mimic rice and wheat in the whole country.

AREA, PRODUCTION AND YIELD TRENDS
OF THE RICE-WHEAT CROPPING SYSTEM
IN THE INDO-GANGETIC PLAINS

Area Trend: Rice

The acreage under rice has been steadily increasing in South Asia from 1958 to 1997. The increase in rice area growth in South Asia has been linear at 0.64% per annum rising from 43 million to more than 58 million hectares in 1998 (Figure 2A). The total rice acreage is highest for India, followed by Bangladesh, Pakistan and Nepal. Total acreage under rice increased from 33 million to over 43 million hectares in India at 0.6% per annum growth between 1958 and 1998 (Figure 2B). In Bangladesh, acreage under rice increased rapidly between 1958 and 1970 from 8 million hectares to over 10 million hectares at 1.9% per annum growth (Figure 2D). But between 1970 and 1997, the increase in the acreage growth under rice has slowed to 0.24% and was around 10.5 million hectares in 1997. The rice area growth increase in Paki-

FIGURE 2. Rice area trends in South Asia, 1958-98.

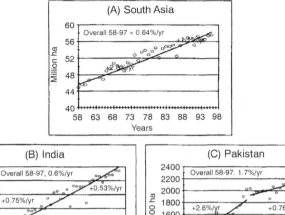

(A) South Asia

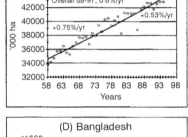

(B) India

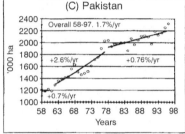

(C) Pakistan

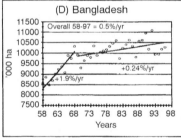

(D) Bangladesh

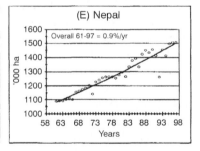

(E) Nepal

stan was initially slow at 0.7% per annum, thereafter increasing rapidly from 1.2 million hectares in 1963 to two million hectares in 1979 at 2.6% annual growth rate. After 1979, the growth in the acreage of rice in Pakistan slowed to a rate of 0.76% per annum (Figure 2C). The total acreage under rice in Pakistan during 1997 was approximately 2.3 million hectares. Overall, the percent annual area growth increase for rice in Pakistan between 1958 to 1997 was 1.7% per annum. The growth in area under rice in Nepal started later in 1961, but this increase has been linear at 0.9% per annum ever since (Figure 2E).

Rice acreage in Nepal increased from 1.1 million hectares in 1961 to above 1.5 million hectares in 1997.

Area Trend: Wheat

Unlike the linear growth increase in area for rice in South Asia, the increase in the acreage of wheat has followed a sigmoid pattern (Figure 3A). Between 1958 and 1967, before the introduction of new varieties, wheat acreage growth for all of South Asia was as low as 0.6% per annum and grew from 17 million to about 18 million hectares in this period. Subsequent increase in wheat acreage growth was rapid at 2.8% per annum and area grew from 20 to more than 32 million hectares between 1967 and 1982. After 1982 area growth of wheat has slowed to 1.1%. Wheat acreage is highest for India, followed by Pakistan, Bangladesh and Nepal. The increase in the wheat acreage in India has followed a sigmoid pattern. Initial growth between 1958 and 1966 was slow at 0.1% annually with 13 million hectares in 1966 (Figure 3B). Between 1966 and 1983, wheat acreage growth rate in India increased at 3% per annum to reach 25 million hectares, after which the growth rate slowed to 1% a year to reach 26 million hectares in 1997. The overall growth rate in India during the period of Green Revolution (1967-1997) was 2%. The wheat acreage growth trend in Pakistan was linear since 1958 at 1.5% per annum (Figure 3C). The acreage rose from a little over 4.5 million hectares in 1958 to around 8.5 million hectares in 1997.

Wheat acreage growth rate in Bangladesh was initially at 3% per annum between 1968 and 1977, then rose to 4% per annum to reach 0.85 million hectares in 1997 (Figure 3D). Overall increase in wheat acreage in Bangladesh was a phenomenal 7.3% per year mainly due to a rapid increase from 1977-1980. Wheat acreage growth rate in Nepal followed a three-step process (Figure 3E). Initial growth rate was stagnant at –0.1% from 1961 to 1966 at 0.1 million hectares. From 1966 to 1986, the growth rate accelerated to 5.8% per annum to reach 0.6 million hectares. Thereafter, the wheat acreage growth rate slowed to 1.3% per annum to reach 0.65 million hectares in 1997. The overall growth rate in Nepal from 1966 to 1993 was an impressive 5.3% per annum.

Area increase under rice and wheat for the entire IGP is slowing as most land becomes double or triple cropped and more land goes for housing, industry, and other purposes. That means that future growth

FIGURE 3. Wheat area trends in South Asia, 1958-98.

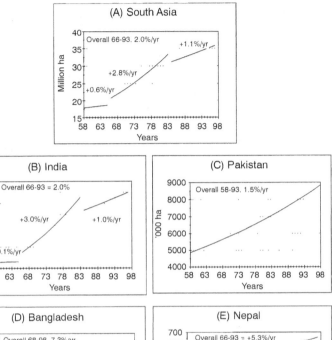

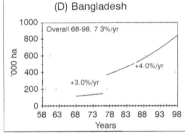

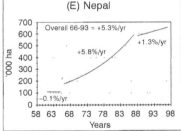

in production will have to come from increase in yield growth rates using technology that ensures sustainable growth and minimal damage to the resource base. Land will become more limiting and growth rates may even become negative.

Production Trend: Rice

Total rice production in South Asia has increased from 60 million tons in 1958 to 160 million tons in 1997 at an impressive growth rate of 2.5% per annum (Figure 4A). Total production has been the highest for India, followed by Bangladesh, Pakistan and Nepal. The production growth rate in India for rice has been generally linear between 1958 and 1997 at 2.6% per annum (Figure 4B). Total production in India was approximately 45 million tons in 1958, which increased to 120 million tons in 1997. The rice production growth trend in Bangladesh has been linear, similar to that of India's at 2.1% per annum (Figure 4D). Production in Bangladesh increased from 10 million tons in 1958 to around 30 million tons in 1997.

The production growth rate in Pakistan has seen a sigmoid growth

FIGURE 4. Rice production trends in South Asia, 1958-98.

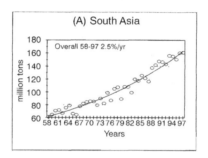

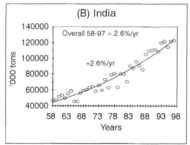

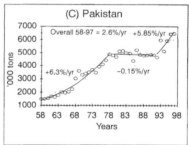

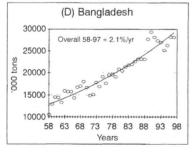

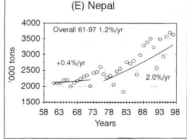

pattern—a growth rate of 6.3% from 1 million tons in 1958 to 5 million tons in 1979 due to the cultivation of modern high yielding varieties (Figure 4C). This impressive initial growth rate slowed to zero between 1979 and 1991 with production staying at 5 million tons, and then grew as production rose to 6.5 million tons between 1991 and 1997 at a growth rate of 5.8% per annum. The stagnant growth rate of rice production between 1979 to 1991 in Pakistan was largely due to cultivation with basmati rice, a traditional long duration, low yielding rice preferred for its aroma and flavor and for export. The overall production growth rate for Pakistan from 1958 to 1997 was 2% per annum. The rate of rice production growth increase started later in Nepal at an initial rate of 0.4% per annum between 1961 and 1973, after which the production growth increased at 2% per annum to reach 3.6 million tons in 1997 due to the introduction and spread of modern high yielding varieties (Figure 4E).

Production Trend: Wheat

The wheat production growth rate in South Asia was very rapid at 4.3% between 1967 and 1997 after an initial slow growth rate of 1% per annum from 1958 and 1967 (Figure 5A). For South Asia, the overall production growth rate since 1958 was an impressive 5.2% per annum for registering 90 million tons of wheat production in 1997. Wheat production has been the highest for India, followed by Pakistan, Bangladesh and Nepal. The overall wheat production in India registered a growth rate of 5.4% after an initial slow growth of 1% per annum between 1958 and 1967, followed by a rapid growth of 4.4% between 1967 and 1997 (Figure 5B). The total production in India reached 70 million tons in 1997. The overall wheat production in Pakistan was 4.2%, the increase being slow between 1958 and 1966 at 1.4%, thereafter increasing rapidly at 3.6% per annum between 1966 and 1998 (Figure 5C). Total wheat production in Pakistan reached 18 million tons in 1997.

Production growth rate increased in Nepal at a phenomenal 6.1% per annum from slightly over 0.1 million tons in 1961 to more than 1 million tons in 1997 (Figure 5E). In Bangladesh, wheat production trend was slow at 0.8% between 1968 and 1974, being stagnant at around 0.1 million tons (Figure 5D). Thereafter, production increased at 21.8% between 1974 and 1983 to register more than 1.2 million tons of wheat. From 1983 to 1992, wheat production in Bangladesh

FIGURE 5. Wheat production trends in South Asia, 1958-98.

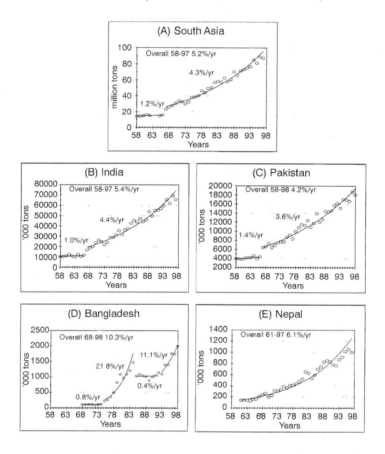

stagnated with only 0.4% growth rate, but, after 1992, wheat production increased again at 11.1% per annum to reach around 2 million tons. The overall wheat production growth rate for Bangladesh was 10.35% per annum. The slack in the production growth in Bangladesh between 1983 and 1992 was largely due to the popularity of growing "Boro" or winter rice, which coincided with the wheat-growing season. Cultivation of "Boro" rice was possible due to development of its new varieties, and the popularity of shallow tubewell installations, which allowed winter rice to be grown in a dry season through assured irrigation. After 1992, the wheat production in Bangladesh increased again for several reasons. First, "Boro" rice required too much water on coarser soils, where wheat is preferred. Second, the introduction of two-wheeled Chinese tractors made it easier to prepare land for wheat and allowed timelier planting and better yields.

Yield Trend: Rice

Overall yield growth rate for rice in South Asia was 1.85% per annum between 1958 and 1997 (Figure 6A). This includes the stagnant period from 1958 to 1967, after which yield growth rate regis-

FIGURE 6. Rice yield trends in South Asia, 1958-98.

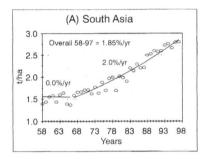

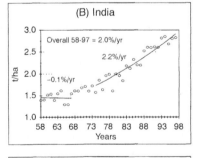

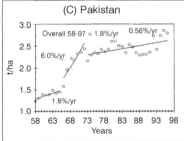

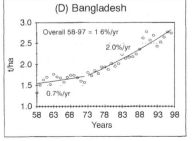

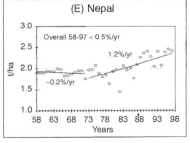

tered an increase of 2% per annum. Mean rice yield of 2.8 tons per hectare in 1997 was double that of the mean rice yields of 1958. The overall yield growth rate in Pakistan for rice was 1.8% per annum (Figure 6C). This increase was characterized by a gradual growth of 1.8% between 1958 and 1967, followed by a very rapid growth between 1967 to 1973 of 6%, decreasing thereafter to 0.56% per annum between 1973 and 1997 as more land was grown to low yielding *Basmati* rice varieties. After an initial stagnating rice yield growth rate between 1958 and 1967 in India, yield increased at 2.2% per annum to reach around 2.8 tons per hectare in 1997 (Figure 6B). Though scented or basmati rice is grown in selected districts of India, its acreage compared to non-scented high yielding modern varieties is only a fraction. Therefore, overall yield growth rate for India has been almost linear at 2% per annum.

The rate of rice yield growth in Nepal hasn't been consistent. The overall growth rate has been 0.5% between 1958 and 1997; this was due to a decline in rice yield from 1958 to 1973, followed by an increase in yield at 1.2% per annum from 1973 and 1997 (Figure 6E). The fluctuating yield trend for this water-loving rice crop in Nepal is a reflection of the yearly variation in the onset and duration of monsoon rains, as assured irrigation sources in Nepal is not as widespread as in the neighboring countries. In Bangladesh, the rate of rice yield increase has been a steady 1.6% per annum (Figure 6D). The initial years saw an increase of only 0.7% between 1958 and 1973, followed by a rapid increase of 2% per annum between 1973 and 1997 to a level of 2.8 tons per hectare. The latter rapid increase in rice yield can be attributed to an rapid increase in Boro rice during this period.

Yield Trend: Wheat

The overall wheat yield growth rate in South Asia has been 3.2% per annum between 1958 and 1997 (Figure 7A). This includes a slower growth of 0.8% between 1958 and 1967 followed by a more rapid 2.8% annual yield growth rate between 1967 and 1997. The average wheat yield in South Asia was around 2.5 tons per hectare in 1997. After a stagnant yield growth rate between 1958 and 1967, wheat yield increased at 2.3% in Pakistan from 0.8 tons in 1967 to 2.2 tons per hectare in 1997 (Figure 7C). Wheat yield increased from 0.8 tons per hectare in 1958 to 2.8 tons per hectare in 1997 in India (Figure 7B). The overall wheat yield growth rate in India was 3.3% per annum.

FIGURE 7. Wheat yield trends in South Asia, 1958-98.

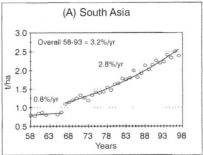

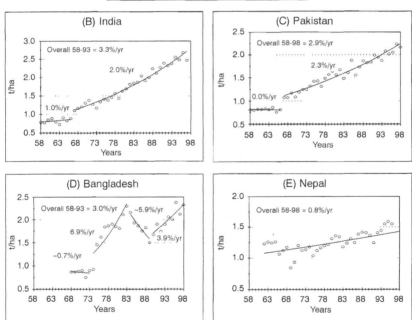

Initially, from 1958 to 1967, yield growth rate increase was slow at 1% per annum, but increased rapidly to 3% per annum between 1967 and 1997 in India.

Wheat yield growth increase in Nepal fluctuated over the years. The overall wheat yield increase in Nepal has been 0.8% per annum from 1962 to 1998 (Figure 7E). The average yield in Nepal was around 1.6 tons per hectare in 1998. The yearly fluctuations in wheat yield for

Nepal is a reflection of late planting of wheat over large areas especially for years registering low yields. Despite an overall yield increase of
3% per annum in Bangladesh between 1958 and 1993, the trend during these years has been very inconsistent (Figure 7D). Yield growth
rate was stagnant during 1968 and 1974, increased rapidly between
1974 and 1983 at 6.9% per annum, and then declined almost at a
similar rate of – 5.9% between 1983 and 1992.

Unlike the rest of the IGP, total area under irrigated rice in Bangladesh increased 100% between 1980 and 1990 due to the popularity of
shallow tube wells. As a result of this single factor and the availability
of newer varieties, acreage under winter "Boro" rice during the post-
monsoon winter season increased by 87% and spring "Aus" rice by
51% during this period. These changes in the rice production scenario
in Bangladesh brought about stagnation in the production and yield of
wheat between 1980 and 1990 since "Boro" rice competed with
wheat for land (Bhuiyan et al., 1993; Razzaque, Badaruddin and Meiss-
ner, 1995). But between 1992 and 1997, the wheat yield increased at
3.9% per annum to around 2.3 tons per hectare in 1997. This post 1992
surge in wheat production and yield of wheat in Bangladesh was
probably due to falling prices of rice, and more cost of production,
specially irrigation, of Boro rice. It was also helped by introduction of
two-wheel Chinese tractors that allowed timelier and easier wheat
planting and higher yields. Thus areas with coarser soils and poor
water retention capacity, which were being cultivated under Boro rice,
were reverted to wheat cultivation (Bhuiyan et al., 1993).

REGIONAL DIAGNOSTIC AND BENCHMARK SURVEYS: CONSTRAINTS OF THE RICE-WHEAT CROPPING SYSTEM

"Sustainability" as a research objective is well recognized but its
appropriate quantitative measurements is not well defined given the
vast array in which it can be characterized and assessed (Harrington et
al., 1992). The total production of rice and wheat in the Indo-Gangetic
region has increased steadily in the last two decades. Under this scenario, it is often difficult to produce evidence of stagnating or declining productivity of the rice-wheat cropping system in the farmer's
field. For this article, we define production as the total quantity of crop
produce of economic value and productivity as the quantity produced
per unit factor of interest. The unit factor of interest could be unit land

area (e.g., yield), or any input. Other factors are also important to be considered but it is more difficult to document with sufficient accuracy the usage of these factors and its quantities in farmer's field, since farmers rely on memory, estimation, and experience rather than well-maintained records in South Asia.

Monitoring agronomic efficiency on a long-term basis within and between farmer's field and agro-climatic zones require a lot of resources and long-term commitment is essential. The quality and recommended quantity of crop inputs, the ease of access to these inputs, crop management and post-harvest storage and processing, economic condition of a farm household, market access and incentives, and prevailing weather conditions form a network of intertwined factors that determine agronomic efficiency. However, a regional effort between the national Institutes of the host countries of the IGP and the International agricultural centers has been to answer the question–Is the productivity of the rice-wheat cropping system stagnating or declining? Answering in the affirmative should be followed by an effort in understanding the underlying cause of yield stagnation or decline and searching for remedial solutions. As per the evidence provided in the subsequent articles, it is clear that within the IGP, solutions towards future yield growth will be eco-regional and site specific.

Three forms of simultaneous activities have been on going to collect evidence from farmer's field for possible rice and wheat yield stagnation or decline in the IGP. They include farmer's field diagnostic and benchmark survey, location specific long-term monitoring (8 to 9 years) surveys and research station managed long-term soil fertility experiments. This section discusses diagnostic and benchmark surveys. Data compilation from long-term monitoring surveys is not complete, and long-term soil fertility experiments suggests that there is evidence for rice and wheat yield decline.

Farmers field diagnostic and long-term monitoring surveys in rice-wheat growing areas of the region were proposed as one way to measure the sustainability and productivity of the rice-wheat cropping system of the IGP (Harrington et al., 1992; Harrington et al., 1993a). In the last decade, several one-time diagnostic and benchmark surveys have been conducted in the Indo-Gangetic region, collecting either qualitative, semi-quantitative or a combination of both forms of data. Since the diagnostic surveys were conducted in different locations, years and crop seasons, and at one time only, it is difficult to combine information from these surveys to analyze multi-locational trends of rice-wheat in farmer's field over time. However, these surveys are a major source of information about the constraints and practices of rice-wheat as it exists in the farmer's field and is therefore very valuable.

This section summarizes information (Table 1) from 15 diagnostic surveys (Ahmed et al., 1993; Baksh and Razzaque, 1999; Byerlee et al., 1984; Fujisaka et al., 1994; Gami et al., 1997; Harrington et al., 1993a; Harrington et al., 1993b; Harrington et al., 1993c; Hobbs et al., 1990; Hobbs et al., 1992; Meisner, 1992; Meisner, 1996; Saunders, 1990; Saunders, 1991; and Singh et al., 1996), and three benchmark surveys (Adhikari et al., 1999; Dhiman et al., 1999; and Hobbs et al., 1996) conducted in this region on rice-wheat in the last 10 years. Multidisciplinary teams of scientists conducted the diagnostic surveys using informal and semi-formal rapid rural appraisal approaches. The surveys lasted from 1 to 2 weeks and featured informal but structured

TABLE 1. List of problems for rice-wheat cropping systems in the Indo-Gangetic Plains based on 15 diagnostic surveys.

Wheat
- Late planting
- Inadequate plant stand
- Weed problems, especially grassy weeds
- Early season waterlogging
- Nutrient deficiencies, both macro- and micro-nutrients
- Late season moisture stress and water management
- Various leaf diseases
- Varietal problems

Rice
- Pests and disease
- Mid-season and late season water stress and water management
- Nutrient deficiencies
- Weeds for direct seeded rice

farmer interviews and covered farmers' practices and knowledge on land types, cropping systems and crops, management, production, field observations and problem identification. Teams met in the afternoon to discuss findings and develop problem and cause diagrams before planning the next day's activity. Most problems were identified by farmers and researchers and prioritized using simple scoring methods. Opportunities for research and extension interventions were identified. A systems rather than a single commodity analysis was done. The three-benchmark surveys were conducted at sites that were chosen for longer term monitoring farmer practices. These surveys provide a detailed description of farms and farmer practices in selected rice-wheat fields.

Characteristics of the Rice-Wheat System

Rice is grown in the Indo-Gangetic Plains in areas with as little as 120 mm rain to areas with more than 1800 mm. Rice is grown in some areas with full irrigation and others with supplemental or just as a rainfed crop. The same is the case for wheat although in the rice-wheat system the majority of fields of wheat get at least one initial irrigation. Drought is therefore an issue in the rice-wheat system where lack of rain can affect the start on the rice season and the yield of the subsequent crops. This can lead to delays in rice planting, but where irrigation is available rice planting can start as early as May (Punjab, India).

Soils and hydrological situations differ throughout the IGP. On the lower terraces with heavier, poorly drained soils wheat is less suited whereas on higher terraces with coarser soils and good drainage, rice is less suitable. In the Eastern IGP two or three rice crops may be grown instead of rice-wheat on the lower terraces. Wheat is then found on higher terraces where there is better drainage and better soils. On coarser textured soils in the Western IGP rice is grown with irrigation and wheat follows rice.

In the rice-wheat systems, rice and wheat are not the only crops grown. Rotations include many substitute crops for wheat in the winter season including oilseeds (brassica, flax, sunflower, etc.), pulses (gram and lentil) and fodders (berseem clover, maize, etc.). In some rainfed areas mixtures of oilseeds and legumes are grown with wheat and in some irrigated areas brassica is commonly mixed with wheat. In some areas, sugarcane is grown in rotation with rice and wheat.

Otherwise there are not many alternatives for rice in the wet monsoon period, except some fodder crops, although in Bangladesh jute can substitute for rice. In the western reaches of the IGP only one crop of rice and one of wheat is possible in a year, but in eastern areas, temperatures are higher and growth duration less and two rice crops and one wheat crop is possible. Or in lower lying areas two or three rice crops are grown in a year.

Rice in the rice-wheat system is normally seeded in late May and June and transplanted in late June through July. Harvest can be as early as October and continue through November, especially where photosensitive local or quality rice is grown (e.g., *Basmati* rice). Wheat can be planted as early as late October, is best planted in mid-November and can be planted late in December or even January. Harvest starts in March in the east and continues into May in the west.

The most common problems identified from the various diagnostic surveys are listed in Table 1. Each of these will be discussed in the following text with some researchable issues and solutions suggested. Some are simply issues for the single commodity programs and are presently being researched. Others have a systems perspective and these will be highlighted here.

One of the major problems of this system is the completely different way that rice and wheat soils are managed. Transplanting rice seedlings into puddled soils is the traditional way to grow rice in this region (soils plowed when wet). There is some dry seeded, non-puddled rice but this is mainly found in rainfed areas. The puddling is done to restrict water percolation and losses and through the water better control the weeds. This is optimal for rice, but does create problems for the following wheat crop that are planted into physically degraded, puddled rice soils. The repeated cycle of wet puddling soils for transplanted rice over many years has led to several problems. These include the deterioration of soil physical structure, formation of a hard pan at a shallow depth, poor infiltration and waterlogging and poor rooting. It is also one of the main causes for late planting of wheat since farmers need to plow their fields many times to get a good seedbed for planting wheat. The lack of soil physical structure produces a cloddy structure when medium to heavy puddled soils is plowed. This excessive tillage and poor soil structure also creates a problem during the first irrigation. Water stands in the plot due to poor infiltration, oxygen stress occurs and wheat plants turns yellow and are stunted and set back in growth.

One solution to this problem is zero-tillage. This is described in another chapter. In this system, wheat is planted into the rice stubble with minimum disturbance. This can be done from merely broadcasting the seed on the surface to use of planting equipment designed to place seed in the soil without tillage. Results are impressive and farmers who adopt this technology are describing a whole array of benefits. These range from cost and fuel saving, to better yield, no yellowing after irrigation, less weeds, less water needed and less lodging. There are also many environmental benefits including less carbon emissions and less global warming, less herbicide and water use. This solution overcomes the problem of late planting and poor plant stands.

Another research tact is to do away with puddling in soils where this is not needed (heavy, poor draining low land and salt affected soils). Weeds become a major issue in this case and weed strategies are needed to overcome this new problem. Use of integrated weed management is a key to solve this problem. Without puddling, soil physical poperties are improved and the subsequent wheat crop, especially with zero-till is much better.

Nutrient imbalance and mining by rice and wheat has led to problems in some areas. Both these cereal crops are heavy users of nutrients. In many areas, most of the crop residues are removed for animal or fuel use, especially in the east. Nitrogen and phosphorus are two of the most important macro-nutrients that become limiting in the rice-wheat system. Of course application of fertilizer, either organic or inorganic or a combination of the two can overcome this problem. Potassium is also starting to become limiting in some rice-wheat areas. This element is usually present in adequate amounts in the IGP soils, but because many farmers remove the entire residue and don't apply potassium, this imbalance is leading to deficiency. Zinc, boron, manganese are micronutrient problems in some areas. Boron leads to sterility in wheat and zinc is essential for rice yield in some areas, especially where there are alkaline soils. Soil organic matter declines have been suggested as a problem with the rice-wheat system. The use of organic fertilizers in combination with inorganic ones shows better results. However, organic manures are in short supply since they are needed for fuel purposes and it is costly to transport to the fields. Some long-term experiment results also indicate that organic matter can increase or stabilize over time in this system (Duxbury et al., 2000).

Solutions to this nutrient problem are found in finding ways to more efficiently utilize applied nutrients. This is also covered in another chapter. Essentially a system is needed to maintain a balance between what is supplied and what is removed. Studies are underway to determine the nutrient supplying capacity of individual soils. Based on this a recommendation can be calculated on how much external nutrients are needed to meet a certain yield level without causing a soil imbalance.

Continuous rice-wheat has led to buildups of pest and diseases of rice and wheat. One of the best examples of this is the spread of the grassy weed, *Phalaris minor*. This is a major problem in NW India, where this weed has developed resistance to the commonly used herbicide, *Isoproturon* ™. Use of new reduced tillage options can help since the soil is not disturbed and weeds germinate less. Also through timely planting, the wheat can emerge and then shade out the later planted weed plants. Use of bed planting can be used to reduce the weed population while reducing herbicide use. New herbicides are also available, but their cost is high and they are also susceptible to herbicide resistant weeds. Rotations with other crops are also a viable integrated strategy to adopt in rice-wheat systems. Stemborer issues in zero-tillage wheat have been raised as a possible problem if these new technologies are adopted. This is being closely monitored but initial data indicates this is less of a problem than anticipated. In fact in zero-tillage where the residues are left on the surface and not buried, beneficial insects have a better habitat and survive to reduce this insect problem. The occurrence of diseases that carryover from rice to wheat and are perpetuated by the system are less than anticipated although they are being monitored, both foliar and soil pathogens.

Issues of water management figure high in this system. Many farmers list water stress as an issue. Water pricing policy is to blame in some areas, since the incorrect policy can lead to misuse of this natural resource. In many areas of NW India and Pakistan water tables are declining rapidly as more water is removed than replaced. This is obviously not a sustainable situation. In other areas water tables are rising and leading to waterlogging or salinity/sodicity. Sound water use policies and good water management strategies are needed. Watershed water balance studies are needed at the macro-level. At the field level, zero-tillage, bed planting, laser leveling and dry seeded rice are being monitored as ways to improve water use efficiency.

Varietal use is another problem in the rice-wheat system. This is especially a problem in areas where late maturing rice varieties are grown. *Basmati* rice is high quality scented rice grown in NW India and Pakistan (80% in Punjab Pakistan). It matures in November and therefore interferes with timely wheat planting. Researchers are working to find shorter duration *Basmati* rice varieties but at the moment farmers must reduce turnaround time using reduced tillage techniques to overcome this problem.

In India, in addition to the agronomic reasons, the yearly declaration of a remunerative support purchase price for rice and wheat by the government has helped the rice-wheat rotation to spread from west to east. This spread has happened in many states in India by displacing a sizeable area under legumes and other less productive crops, e.g., oilseeds and millets. Obviously policy is a major issue in rice-wheat systems.

A majority of the farmers save their own seeds for planting the following year. Farmers cite this as a factor that limits optimum plant population. This is more of a problem for wheat because wheat seeds need to be stored through the hot and humid summer months and then directly seeded at the time of planting. Rice seeds are stored through the cooler and less humid winter months and secondly, only germinated and healthy rice seedlings are transplanted, therefore, rice seed quality for transplanted rice is less of a concern than wheat seed quality.

It is ironic that in a region that supports more than a billion people, labor shortages during the major cultural operations, i.e., at planting, harvesting and threshing is increasing. Renting of farm equipments in the northwestern IGP and the adoption of small-scale mechanization in the eastern IGP is increasing. In the northwestern IGP, renting of tractors for land preparation, and combine harvesters for harvesting rice and wheat is common. The use of locally manufactured threshers for wheat and rice is more common in the western IGP compared to the eastern IGP. It is envisaged that with the introduction and spread of two-wheeled hand operated tractors in the eastern IGP for the major cultural operations, labor shortages will be overcome to some extent. Though substantial improvements have been made in the availability of credit for farmers, farmers still cite the easy and timely access to credit and effective marketing of their produce as a major limitation in their overall farming needs.

PRODUCTIVITY OF THE RICE-WHEAT CROPPING SYSTEM

The total production of rice and wheat in South Asia has been increasing on a yearly basis, but there are concerns that the productivity of this system is either stagnating or declining. Productivity trend analysis can be done on a regional scale based on published government statistics, at the farm level to understand changes in farmers field through long-term farm monitoring, or on research stations based on the data of long-term experiments. This section summarizes results from reports of productivity analysis done on data reported in government publications (Chand, 1999; Chand and Haque, 1998; Hobbs and Morris, 1996; Indian Council of Agricultural Research, 1998; Kumar et al., 1998).

Changes in the yield trend of a cropping system should be studied in terms of the inputs used, to make a realistic assessment of the productivity of a cropping system. Few total (TFP) and partial factor productivity (PFP) analyses have been done for the rice-wheat cropping system of South Asia as a measure of sustainability. A positive trend in the factor productivity over several years of a cropping system would indicate increases in the quantity and quality of input used, and favorable changes in the technological, physical and economic environment (Harrington et al., 1992). Hobbs and Morris (1996) have appropriately summarized the factor productivity studies done on rice-wheat cropping system prior to 1996. The few TFP studies on rice-wheat cropping system indicate a declining or a negative productivity for the system (Ali and Velasco, 1994; Cassman and Pingali, 1995) due to deterioration of resources, and changes in the physical and chemical properties of the soil thereby reducing nitrogen-supply capacity of the soil. PFP studies also indicate a decline in the productivity of the rice-wheat system (Byerlee and Siddique, 1994; Kumar and Rosegrant, 1994). The TFP analyses done by Kumar et al. (1998) shows a declining productivity in the IGP region of India. The authors argue that the yield growth has been more input based, and higher growth in yield and production in the IGP is only possible through better management of existing soil and water resources. Pingali and Heisey (1999) have also made similar conclusions during their analyses of cereal crop productivity of developing countries. Therefore, in intensively cultivated areas where two or more cereal crops are grown in a rotation, there is evidence of finite resource base limiting productivity growth (Pingali and Heisey, 1999), and its growth can only be sustained by conserving the natural resource base, increasing input use efficiency, and by increasing the investments in agricultural research and education.

CONCLUSIONS

To conclude, the rice-wheat systems are here to stay. They are a major source of cereal food production in the region and it will be difficult to replace them. Farmers find very few substitutes for rice and wheat that provide similar low risk and profit. Research has to find ways of sustaining this system and making it more efficient and profitable. Continued growth in rice and wheat area is very unlikely since in the future there will be more competition for domestic and industrial uses. This leaves just yield growth available to increase production growth. The use of cost reducing, more efficient systems will be required if agriculture in the region is to keep pace with population demand and food security is to be maintained.

REFERENCES

Adhikari, C.A., C.B. Adhikary, N.P. Rajbhandari, M. Hooper, H.K. Upreti, B.K. Gyawali, N.K. Rajbhandari, and P.R. Hobbs. (1999). *Wheat and Rice in the Mid-Hills of Nepal: A Benchmark Report on Farm Resources and Production Practices in Kavre District. Kathmandu*: NARC and CIMMYT, pp. 1-26.

Ahmed, N.U., M.M. Alam, A.M. Bhuiyan, R. Islam, and M.U. Ghani. (1993). *Rice Wheat Diagnostic Survey in Greater Kushtia District of Bangladesh*. Rice Farming Systems Division, Bangladesh Rice Research Institute, Gazipur-1701, Bangladesh, pp. 1-57.

Ali, M., and L.E. Velasco. (1994). *Intensification of Induced Resource Degradation: The Crop Production Sector in Pakistan*. Los Banos, Phillipines: International Rice Research Institute (IRRI). Mimeo.

Baksh, Md.E., and M.A. Razzaue. (1999). *Diagnostic Survey Report on Rice-Wheat system at Dinajpur and Rangpur Districts*. Wheat Research Centre, Bangladesh Agricultural Research Institute, Nashipur, Dinajpur, pp. 1-38.

Bhuiyan, A.M., M. Badaruddin, N.U. Ahmed, and M.A. Razzaque. (1993). *Rice-Wheat System Research in Bangladesh: A Review*. Bangladesh Rice Research Institute and Bangladesh Agricultural Research Institute, pp. 1-96.

Biswas, T.D., and G. Narayanaswamy. (1997). Sustainable Soil Productivity Under Rice-Wheat System. *Indian Society of Soil Science* Bulletin No. 18, pp. 1-83.

Byerlee, D., A.D. Sheikh, M. Aslam, and P.R. Hobbs. (1984). *Wheat in the Rice-Based Farming System of the Punjab: Implications for Research and Extension.* AERU, NARC and CIMMYT, Pakistan.

Byerlee, D., and A. Siddique. (1994). Has the Green Revolution been sustained? The quantitative impact of the seed-fertilizer revolution in Pakistan revisited. *World Development.* 22(9): pp. 1345-1361.

Cassman, K.G., and P.L. Pingali. (1995). Extrapolating trends from long-term experiments to farmers' fields: The case of irrigated rice systems in Asia. pp. 63-84. In *Agricultural Sustainability: Economic, Environmental, and Statistical Considerations,* eds. V. Barnett, R. Payne, and R. Steiner, London, U.K.: John Wiley,

Chand R. (1999). Emerging Crisis in Punjab agriculture: Severity and options for future. *Economic and Political Weekly.* March 27, pp. A2-10.

Chand R., and T. Haque. (1998). Rice-wheat crop system in Indo-Gangetic Region: Issues concerning sustainability. *Economic and Political Weekly.* Vol 33 (29): pp. A108-12.

Dhiman, S.D., M.K. Chaudhary, D. Singh, H. Om, D.P. Mandal, D. Singh, and D.V.S. Panwar. (1999). *Measuring Sustainability of Rice-Wheat Cropping System Through On-Farm Monitoring in Haryana.* CCS Haryana Agricultural University, Rice Research Station, Kaul, Regional Research Station, Karnal publication, p. 37.

Duxbury, J.M., I.P. Abrol, R.K. Gupta, and K.F. Bronson. (2000). Analysis of soil fertility experiments with rice-wheat rotations in South Asia. In *Long Term Soil Fertility Experiments in Rice-Wheat Cropping Systems,* eds. I.P. Abrol, K.F. Bronson, J.M. Duxbury and R.K. Gupta. Rice-wheat consortium paper series number 6. New Delhi, India. RWC for IGP.

Fujisaka, S., L.W. Harrington and P.R. Hobbs. (1994). Rice-wheat in South Asia: Systems and long-term priorities established through diagnostic research. *Agricultural Systems* 46: 169-187.

Gami, S.K., B. Chaudhary, and M.P. Shah. (1997). *Rice-Wheat Diagnostic Survey in Rauthat District of Nepal.* Nepal Agricultural Research Council and SM-CRSP Cornell University, USA, pp. 1-39.

Harrington, L.W., S. Fujisaka, P.R. Hobbs, C. Adhikary, G.S. Giri, and K. Cassaday. (1993a). *Rice-Wheat Cropping System in Rupandehi District of the Nepal Terai: Diagnostic Surveys of Farmers' Practices and Problems, and Needs for Further Research,* Mexico, D.F.: CIMMYT, NARC, and IRRI, pp. 1-33.

Harrington, L.W., P.R. Hobbs, K.A. Cassaday, eds. (1992). *Methods of Measuring Sustainability Through Farmer Monitoring: Application to the Rice-Wheat Cropping Pattern in South Asia.* Proceedings of the workshop, 6-9 May, 1992, Kathmandu, Nepal. Mexico, D.F.: *CIMMYT, IRRI and NARC,* pp. 1-111.

Harrington, L.W., P.R. Hobbs, D.B. Tamang, C. Adhikari, B.K. Gyawali, G. Pradhan, B.K. Batsa, J.D. Ranjit, M. Ruckstuhl, Y.G. Khadka, and M.L. Baidya. (1993b). *Wheat and Rice in the Hills: Farming Systems, Production Techniques and Research Issues for Rice-Wheat Cropping Patterns in the Mid-Hills of Nepal.* Report on an exploratory survey conducted in Kabhre district, 20-28 April 1992. *NARC CIMMYT.* Mexico DF. 84 p.

Harrington, L.W., S. Fujisaka, M.L. Morris, P.R. Hobbs, H.C. Sharma, R.P. Singh, M.K. Chaudhary, and S.D. Dhiman. (1993c). *Wheat and Rice in Karnal and Kurukshetra Districts, Haryana, India: Farmers' Practices, Problems, and an Agenda for Action.* Mexico D.F. HAU, ICAR, CIMMYT and IRRI.

Hobbs, P.R., G.P. Hettel, R.P. Singh, Y. Singh, L.W. Harrington, and S. Fujisaka. (1990). *Rice-Wheat Cropping Systems in the Tarai Areas of Nainital, Rampur, and Pilibhit Districts in Uttar Pradesh, India: Diagnostic Surveys of Farmers' Practices, Problems and Needs for Further Research.* Mexico D.F. ICAR, GBPUAT, CIMMYT and IRRI.

Hobbs, P.R., G.P. Hettel, R.K. Singh, R.P. Singh, L.W. Harrington, V.P. Singh and K.G. Pillai. (1992). *Rice-Wheat Cropping Systems in Faizabad District of Uttar Pradesh, India: Diagnostic Surveys of Farmers' Practices, Problems and Needs for Further Research.* Mexico D.F. ICAR, NDUAT, CIMMYT and IRRI.

Hobbs, P.R., L.W. Harrington, C.A. Adhikary, G.S. Giri, S.R. Upadhyay, and B. Adhikary. (1996). *Wheat and Rice in the Nepal Tarai: Farm Resources and Production Practices in Rupandehi District.* Mexico, D.F.: NARC and CIMMYT, pp. 1-44.

Hobbs, P., and M. Morris. (1996). *Meeting South Asia's Future Food Requirement from Rice-Wheat Cropping Systems: Priority Issues Facing Researchers in the Post-Green Revolution Era.* NRG paper 96-01. Mexico, D.F.: CIMMYT, pp. 1-46.

Inayatullah, C., Ehsan-Ul-Haque, Ata-Ul-Mohsin, A. Rehman, and P.R. Hobbs. (1989). *Management of Rice Stem Borers and the Feasibility of Adopting No-Tillage in Wheat.* Entomological Research Laboratories, National Agricultural Research Centre, Pakistan Agricultural Research Council, Islamabad, pp. 1-64.

Indian Council of Agricultural Research. (1998). *Decline in Crop Productivity in Haryana and Punjab: Myth or Reality.* Report of the Fact Finding Committee May 1998, pp. 1-90.

Kumar, P., and M.W. Rosegrant. (1994). Productivity and sources of growth for rice in India. *Economic and Political Weekly.* 29(53): A183-A188.

Kumar, P., P.K. Joshi, C. Johansen, and M. Asokan. (1998). Sustainability of Rice-Wheat Based Cropping Systems in India: Socio-Economic and Policy Issues. *Economic and Political Weekly.* (26): pp. A152-58.

Meisner, C.A. (1992). *Report of an On-Farm Survey of the Mymensingh Region and Tangail Wheat Growers' Practices, Perceptions, and Their Implications. Monograph No. 9.* Bangladesh Agricultural Research Institute, Wheat Research Centre, Nashipur, Dinajpur, pp. 1-31.

Meisner, C.A. (1996). *Report of an On-Farm Survey of the Greater Comilla Region Wheat Growers' Practices, Perceptions, and Their Implications. Monograph No. 13.* Bangladesh Agricultural Research Institute, Wheat Research Centre, Nashipur, Dinajpur, pp. 1-34.

Paroda, R.S., T. Woodhead, and R.B. Singh. (1994). *Sustainability of Rice-Wheat Production Systems in Asia,* RAPA Publication 1994/11, pp. 1-209.

Pingali, P.L., and P.W. Heisey. (1999). *Cereal Crop Productivity in Developing Countries.* CIMMYT Economics Papers 99-03. Mexico D.F.: CIMMYT. pp. 1-32.

Ranjit, J.D., N.K. Rajbhandari, R. Bellinder, and P. Kataki. (1999). *Mapping Phalaris minor in the Rice-Wheat Cropping System of Different Agro-Ecological Regions of Nepal.* Nepal Agricultural Research Council, Nepal and SM-CRSP Cornell University, USA, pp. 1-65.

Razzaque, M.A., M. Badaruddin, C.A. Meisner, eds. (1995). *Sustainability of Rice-Wheat Systems in Bangladesh. Proceedings of the Workshop.* Bangladesh Agricultural Research Institute, Bangladesh Rice Research Institute and Bangladesh Australia Wheat Improvement Project, pp. 1-113.

Saunders, D.A. (1990). *Report of an On-Farm Survey Dinajpur District Wheat Farmers' Practices and Perceptions. Monograph No. 6.* Bangladesh Agricultural Research Institute, Wheat Research Centre, Nashipur, Dinajpur, pp. 1-38.

Saunders, D.A. (1991). *Report of an On-Farm Survey Jessore and Kushtia Wheat Farmers' Practices and Perceptions. Monograph No. 8.* Bangladesh Agricultural Research Institute, Wheat Research Centre, Nashipur, Dinajpur, pp. 1-30.

Singh, R.A., R.M. Singh, H.P. Agrawal, S. Nagarajan. (1996). *Rice-Wheat Cropping System in Varanasi District, Uttar Pradesh, India,* Research Bulletin No. 3, Directorate of Wheat Research, Karnal, India, pp. 1-116.

Yadav, R.L., K.S. Gangwar, and K. Prasad. (1998). *Dynamics of Rice-Wheat Cropping System in India.* Technical Bulletin, Modipuram, Meerut, India, Project Directorate for Cropping Systems Research, pp. 1-90.

Yadav, R.L., K. Prasad, and K.S. Gangwar. (1998). *Analysis of Eco-Regional Production Constraints in Rice-Wheat Cropping Systems Research,* Modipuram, Meerut, India, Project Directorate for Cropping Systems Research, pp. 1-68.

Yadav, R.L., K. Prasad, and A.K. Singh (eds.). (1998). *Predominant Cropping Systems of India: Technologies and Strategies.* Modipuram, Meerut, India, Project Directorate for Cropping Systems Research, pp. 1-237.

Long-Term Yield Trends
in the Rice-Wheat Cropping System:
Results from Experiments
and Northwest India

J. M. Duxbury

SUMMARY. Recent reports have questioned the long-term sustainability of the rice-wheat cropping system. Evaluation of yield trends in fifteen long-term rice-wheat experiments found that rice yields usually declined with time but that wheat yields were more stable. Nutrient response curves indicated that insufficient N was added to achieve maximum yield in some experiments, suggesting that crop lodging prevent genetic yield potential from being reached. Substantial decreases in response of rice to N were observed in some experiments, indicating constraints to productivity other than N supply. Organic inputs used in integrated nutrient management strategies generally substituted for fertilizer nutrients but occasionally increased yield potential. Neither organic inputs nor increased soil organic matter contents markedly affected the sustainability of the rice-wheat system. Wheat yields in the northwest Indian States of Haryana and Punjab increased linearly between 1965 and 1998 and are the major reason for increased wheat production in these States. In contrast, rice yields follow a curvilinear pattern, have stagnated in many districts, and are showing signs of

---

J. M. Duxbury is Professor of Soil Science, 904 Bradfield Hall, Department of Crop and Soil Sciences, Cornell University, Ithaca, NY 14853 USA (E-mail: jmd17@cornell.edu).

Research supported in part by a grant from US AID through the Soil Management CRSP.

[Haworth co-indexing entry note]: "Long-Term Yield Trends in the Rice-Wheat Cropping System: Results from Experiments and Northwest India." Duxbury, J. M. Co-published simultaneously in *Journal of Crop Production* (Food Products Press, an imprint of The Haworth Press, Inc.) Vol. 3, No. 2(#6), 2001, pp. 27-52; and: *The Rice-Wheat Cropping System of South Asia: Trends, Constraints, Productivity and Policy* (ed: Palit K. Kataki) Food Products Press, an imprint of The Haworth Press, Inc., 2001, pp. 27-52. Single or multiple copies of this article are available for a fee from The Haworth Document Delivery Service [1-800-342-9678, 9:00 a.m. - 5:00 p.m. (EST). E-mail address: getinfo@haworthpressinc.com].

possible decline. Increased area in production has been the major factor behind increased production of rice in Punjab State since about 1975. *[Article copies available for a fee from The Haworth Document Delivery Service: 1-800-342-9678. E-mail address: <getinfo@haworthpressinc.com> Website: <http://www.HaworthPress.com> © 2001 by The Haworth Press, Inc. All rights reserved.]*

KEYWORDS. Crop productivity trends, integrated nutrient management, nutrient responses, yield sustainability

INTRODUCTION

Suggestions of stagnating productivity in the rice-wheat rotation of South Asian countries, declining yields in rice-wheat experiments and declining factor productivity have raised questions about the sustainability of this cropping system (Hobbs and Morris, 1996). Yield trends of rice and wheat in fifteen long-term soil fertility management experiments in India and Nepal that had been carried out for eight years or longer are analyzed in this paper. The experiments are spread across the Indo-Gangetic plain (Figure 1) and include soil types ranging in texture from sandy loam to clay loam (Table 1). Two sites in Haryana, India had alkali soils that were ameliorated with gypsum additions.

The contribution of macronutrient nutrition to observed yield trends is evaluated through nutrient response curves. The effects of organic inputs on system productivity and sustainability are assessed in experiments that include integrated nutrient management treatments, and using data from the All India Coordinated Research Project on cropping systems. Results from the experiments are compared with productivity trends in Haryana and Punjab, two States in northwest India with the highest system yields.

CROP YIELD TRENDS IN LONG-TERM EXPERIMENTS

Data were taken from publications on long-term rice-wheat experiments in India (Abrol et al., 2000; Yadav et al., 1998) and Nepal (Nepal Agricultural Research Council, 1998). Evaluation of yield trends in long-term experiments is complicated by considerable year to year

FIGURE 1. Locations of long-term rice-wheat experiments.

Location of Long-Term Rice-Wheat Experiments

N

Ludhiana
Kaipur
Pantnagar
Masaitgaye
Bharatpur
Parwarpur
Ghaghat
Masootha
Jaipur
Faizabad
Pusa
Samastpur
Mohanpur
Bharackpur

Bhutan

Rivers
• Research Sites
Countries
    Bangladesh
    India
    Nepal

0   500   1000 Kilometers

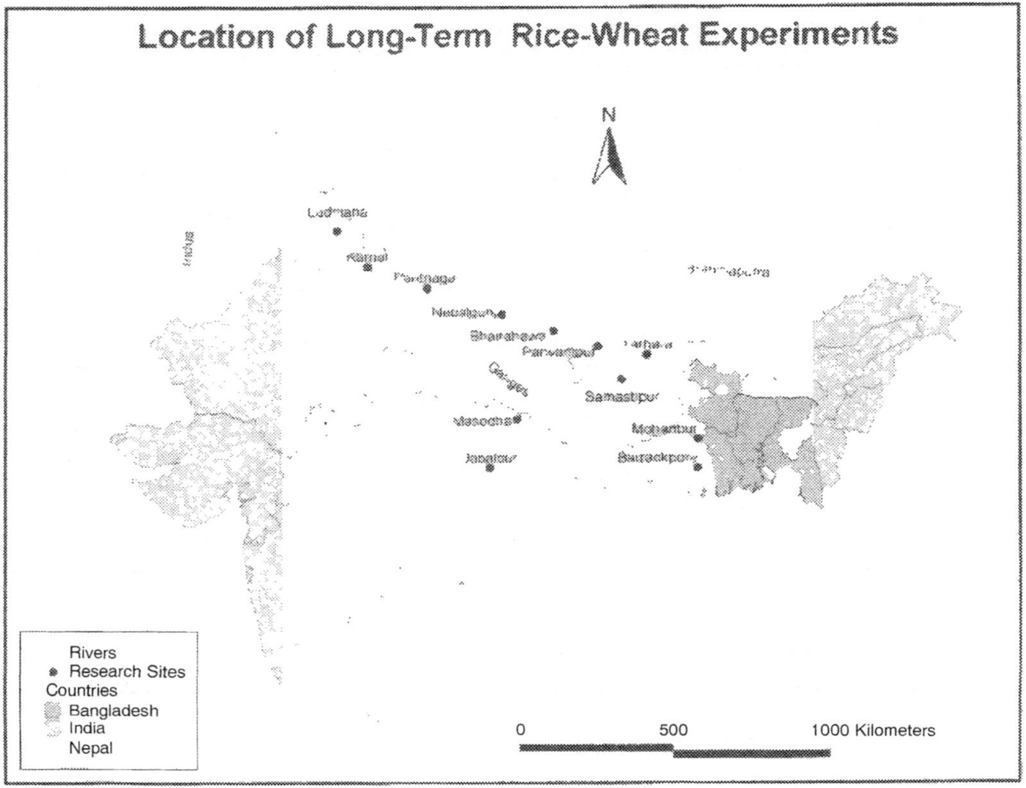

TABLE 1. Site-author key and selected soil properties.

Site	Author(s)	Paper Source[1]	pH[2]	Texture	% O.C.[3] I	% O.C.[3] F	Classification
India							
Ludhiana	Singh, Yd. et al.	A	7.6	loamy sand	0.36	–	Typic Ustochrept
Karnal	Chabbra & Thakur	A	10.3 to 8.5	loam	–	–	Aquic Natrustalf
Karnal	Singh & Swarup	A	9.2 to 8.2	sandy clay loam	0.3	–	Aquic Natrustalf
Pantnagar	Singh, Y. et al.	A	7.0	sandy loam	0.99	0.77	Fluventic Haplaquoll
Pantnagar	Ram	A	7.3	silty clay loam	1.48	0.84	Aquic Hapludoll
Masodha	Yadav R.L. et al.	Ref. list	7.7	silt loam	0.45	0.65	Inceptisol
Jabalpur	Singh and Khan	A	7.3	silty clay loam	0.22	–	–
Samastipur	Sakal	A	8.8	–	0.45	–	Entisol
Mohanpur	Kundu & Samui	A	7.2	sandy loam	0.85	0.72	–
Barrackpore	Saha et al.	A	7.1	sandy loam	0.45	0.41	Euotrochrept
Nepal							
Nepalgunj	Bhattarai & Mishra	B	6.5	silt loam	0.44	0.78	Inceptisol
Bhairahawa	Regni et al.	B	8.0	silt loam	1.03	–	Typic Haplaquept
Parwanipur	Gami & Sah	B	7.0	silt loam	–	0.67	Inceptisol
Tarahara	Yadav, C.R. et al.	B	6.9	loam	–	–	Inceptisol

[1] A is for papers published in Abrol et al., 2000; and B is for papers published in Nepal Agricultural Research Council, 1998.
[2] Values are for baseline soils; where two values are given they represent the baseline and after amelioration with gypsum.
[3] I and F are initial and final values, respectively. Final values are for recommended NPK treatment.

fluctuations in crop yields. Causes for this temporal variability include differences in planting date, weather and pressures from pests and diseases. Cultivars were also changed in many of the experiments but it is assumed that the best available germplasm was always used; that is yields would have been lower had the original cultivars been used throughout the experiments. In order to improve assessment of temporal trends yield data were smoothed by plotting as a 3-year moving average. The one exception to this is an experiment in West Bengal (Saha et al., 2000) where 5-year means were used.

Figures 2 and 3 show temporal trends in crop yields for the highest yielding treatment in each experiment regardless of nutrient sources. In all experiments, one combination of treatments gave the best yields for both rice and wheat. The linear regression model described the yield trends of rice and wheat reasonably well, except for the experiment at Bhairahawa, Nepal where complex yield patterns were observed for the two rice crops in the triple crop rice-rice-wheat rotation. However, use of the linear regression model is not meant to imply that yields will either rise indefinitely or drop to zero. Clearly yield ceilings will be reached where yields are rising and a low level of production can probably be maintained at many sites without fertilizer inputs.

Yields of rough rice declined over time in eight of fourteen experiments (excluding Bhairahawa data), were relatively stable in four experiments, and showed a modest increase in two experiments. In contrast, yields of wheat were more stable over time. Clear declines in yield were observed in only four of the fifteen experiments. Yields were stable in seven experiments and increased in four experiments. Wheat yields were generally lower than rice yields at the beginning of the experiments but were comparable at the end of the experiments, mainly due to declines in rice productivity. Rice-wheat system yield trends were dominated by the trend in rice yields, declining in eight experiments and increasing in two experiments.

The rates of yield change obtained from the linear regression models ranged from -0.48 to $+0.16$ t/ha/yr for rice, from -0.09 to $+0.12$ t/ha/yr for wheat and from -0.56 to $+0.18$ t/ha/yr for the rice-wheat system (Figure 4). Similar yield trends, namely decreasing rice yields with stable or increasing wheat yields, have been reported for three other long-term experiments in India not included in the current analysis (Nambiar, 1994) and these data are also shown in Figure 4. Yield changes were independent of initial yields which ranged from 3.1 to

FIGURE 2. Yield trends for highest yielding treatment (shown at bottom of each panel) in long-term rice-wheat experiments in India. Data are 3-yr moving average, except for Saha et al. where 5-year averages are used.

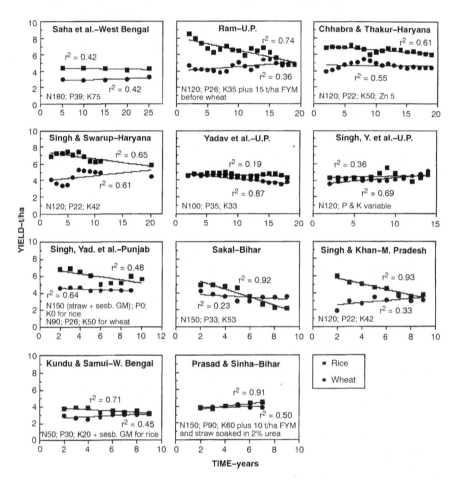

8.5 t/ha for rice, from 1.9 to 4.7 t/ha for wheat and from 5.1 to 13.3 t/ha for the rice-wheat system (average of first three years). The variation in initial yields most probably reflects a combination of differences in site fertility, biotic stresses and crop yield potentials. It is surprising that rice yields are less stable than wheat yields because the paddy environment is considered to be optimal for rice and leads to a poor soil physical condition for growth of wheat, especially in heavier

FIGURE 3. Yield trends for highest yielding treatment (shown at bottom of each panel) in long-term rice-wheat experiments in Nepal. Data are 3-yr moving average.

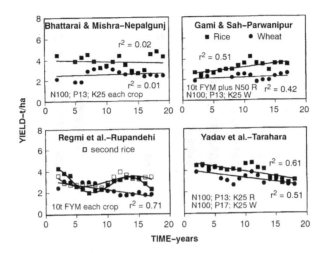

FIGURE 4. Relationship between initial crop yield and average yearly yield change for long-term rice-wheat experiments.

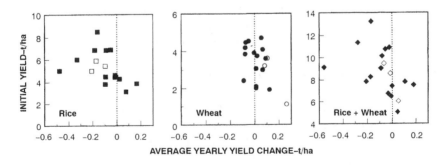

textured soils. Although wheat yields are more stable, they probably are still lower than potential yields.

Temporal changes in rice and wheat yields without nutrient inputs are summarized in Figure 5. Initial yields for rice ranged from 1.6 to 5.8 t/ha, indicating that soil fertility levels and/or other constraints to crop productivity varied considerably across sites at the beginning of the experiments. With two exceptions (Yadav C. R. et al., 1999; Prasad

FIGURE 5. Average yields of rice and wheat without fertilizer addition for the initial and final three years of the long-term rice-wheat experiments.

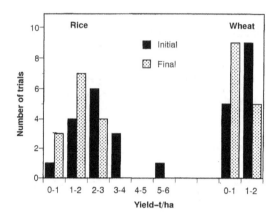

and Sinha, 2000), the final yields of rice were much lower, ranging from 0 to 2.7 t/ha. The mean yield of rice dropped from 2.8 to 1.6 t/ha, a 45% decrease. Wheat yields were again more stable and lower than rice yields. Initial yields ranged from 0.7 to 2.0 t/ha and final yields ranged from 0.2 to 1.6 t/ha. The mean yield for wheat dropped from 1.2 to 0.9 t/ha, a 21% decrease. The most dramatic change occurred at Bhairahawa, in the Nepal terai, where severe P deficiency caused yields to drop to zero for rice and to 0.2 t/ha for wheat.

Crop Yield Response to Nutrient Inputs

Decreases in rice and wheat yields over time are expected in unfertilized treatments as nutrient deficiencies intensify due to crop removal of nutrients without replenishment. Similarly, if soil fertility is not maintained with intensive cropping, nutrient inputs needed to sustain high crop yields will also eventually increase. This scenario may apply to these long-term experiments because nutrient inputs were held constant over time. Yield response patterns to macronutrient inputs provide a means of evaluating whether deficiencies of these nutrients are the cause of yield declines.

Crop yield response patterns to nitrogen inputs for the initial and final phases of the eleven experiments where treatment combinations allow this assessment are shown in Figure 6. Data were averaged for

FIGURE 6. Crop yield responses to nitrogen inputs for the initial and final three years of long-term rice-wheat experiments.

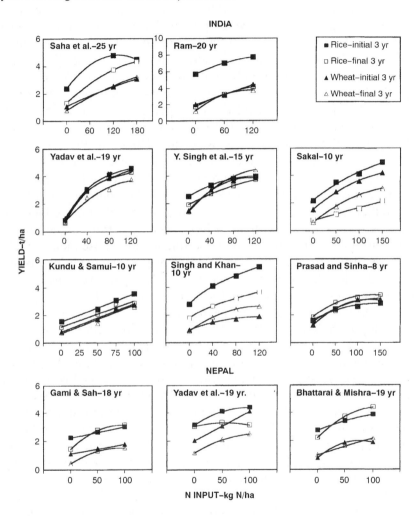

the initial three years and for the final three years of the experiments in order to provide the most representative picture. In five of the eleven experiments (Saha et al., 2000; Ram, 2000; Singh, Y. et al., 2000; Singh and Khan, 2000; and Yadav, R.L. et al., 1998) the yield response has been isolated to a response to N by using data from treatments where P and K were demonstrated not to be limiting yields. Responses to P and K were observed for both rice and wheat in the experiment of Gami and Sah and for wheat in the experiments of Yadav, C.R. et al., 1999 and Bhattarai and Mishra, 1999, and consequently it is uncertain that inputs of these nutrients were optimal. Treatments in three experiments (Sakal, 2000; Kundu and Samui, 2000; and Prasad and Sinha, 2000) were based on increasing additions of N, P and K so that responses cannot be definitively associated with particular nutrients.

In the initial three years of the experiments, a yield plateau for rice is approached in only six of the eleven experiments (Saha et al., 2000; Ram, 2000; Singh, Y. et al., 2000; Prasad and Sinha, 2000; Yadav, R.L. et al., 1998; and Yadav, C.R. et al., 1999). In the remaining experiments, the response patterns clearly show that insufficient nutrients were added to reach maximum yield. The choice not to use higher N inputs may well reflect researcher knowledge that lodging of the crop, rather than genetic potential, becomes the yield-limiting factor at higher N levels. Yields without N input were generally higher at the beginning than at the end of the experiments and consequently yield response curves for the initial and final phases of the experiments are separated. The classical result that higher N inputs were needed to sustain productivity is evident for three experiments where yields were relatively stable over time (Saha et al., 2000; Yadav, R.L. et al., 1998, and Singh, Y. et al., 2000). The response to nutrient inputs actually improved in two experiments where yield was relatively stable (Bhattarai and Mishra, 1999) or increased over time (Prasad and Sinha, 2000). However, in four experiments where rice yields declined over time (Ram, 2000; Singh and Khan, 2000; Sakal, 2000; and Yadav, C.R. et al., 1999), the yield plateau has shifted substantially downwards, indicating the development of a major constraint(s) to crop productivity that is not factored into the experimental design. This constraint(s) could be a nutrient other than N (and most likely also P or K) or some non-nutrient factor(s) such as the build up of root pathogens.

A similar drop in rice yield without N input and shifts in N response curves have been reported for the long-term, continuous double and triple cropped rice experiments at IRRI (Cassman et al., 1996). However, there are important differences between continuous rice experiments and rice-wheat system experiments. At IRRI, soil N levels did not change over time but N availability decreased considerably. This result is explained by changes in the organic forms of N; creation of heterocyclic N compounds through the reaction of phenols with ammonium or amino-containing organics under the more continuous anaerobic environment created in double and triple cropped rice systems (Olk et al., 1998; Olk, Brunetti and Senesi, 1999). The drop in N availability in the rice-wheat system is caused by the commonly observed decline in soil organic carbon and nitrogen levels with cultivation of soils (see following section). Presumably, any stabilization of N during the flooded rice phase is reversed during the wheat phase so that soil organic carbon and nitrogen levels in rice-wheat systems are similar to those in upland cropping systems.

In contrast to rice, wheat yields in control treatments were unchanged over time in eight of the eleven experiments. Yield response curves in the initial and final phases were essentially identical in five of these eight experiments (Saha et al., 2000; Ram, 2000; Kundu and Samui, 2000; Prasad and Sinha, 2000; and Bhattarai and Mishra, 1999). In two experiments (Singh, Y. et al., 2000; and Singh and Khan, 2000), there was a greater response to nutrient additions and higher yields (increase of 0.6 and 1 t/ha, respectively) in the last three years of the experiment. Higher N input was needed to sustain yield in only one experiment (Gami and Sah, 1999). The yield plateau was shifted downwards in three experiments (Sakal, 2000; Yadav, R.L. et al., 1998; and Yadav, C.R. et al., 1999) and yield trends are also downwards in these experiments. Similar to the result with rice, the response of wheat to nutrient inputs was linear in the experiment of Kundu and Samui, 2000, indicating a need for higher nutrient inputs.

Overall, yield response patterns are consistent with the yield trend data and again show that rice is the more vulnerable crop in the rice-wheat rotation. The hypothesis that greater inputs of N, P and K would sustain crop yields as these long-term experiments progressed is supported in only three cases for rice (all requiring greater N inputs) and only one case for wheat. Yield response to N consistently dropped in those experiments where rice and wheat yields declined over time indicating the development of some other constraint. Although response to P inputs was not determined in all experiments, Olsen P values were above the critical level in all experiments so it is unlikely that P deficiency was a causal factor. Deficiency of K could be a contributing factor in some experiments and it is noteworthy that responses to K are increasingly being observed on farms in the Terai of Nepal (J. Tripathi and S.K. Gami, personal communication). However, other nutrient deficiencies, such as S and micronutrients may be limiting the response to macronutrient inputs. The possibility that non-nutrient factors could be constraining productivity also warrants serious consideration.

Effects of Organic Inputs on Crop Yields

Organic inputs were used in eight of the fifteen long-term experiments. *Sesbania aculeata* was grown as a legume green manure (LGM) between wheat and rice in two (Singh, Yd. et al., 2000; and Kundu and Samui, 2000) of the eleven experiments in India. Six experiments included a farmyard manure (FYM) treatment; two in India (Ram, 2000; and Prasad and Sinha, 2000) and all four in Nepal. Three experiments included residue management variables; two in India (Singh, Yd. et al., 2000; and Prasad and Sinha, 2000) and one in Nepal (Gami and Sah, 1999). The best yields of rice and wheat were obtained with an organic input in six of the eight experiments where organic inputs were used (Figures 2 and 3), indicating that these materials play an important role in maximizing crop yields in the rice-wheat system.

In the majority of the experiments, LGMs and FYM were added as substitutes for commercial fertilizers in an integrated inorganic-organic nutrient management approach rather than in addition to fertilizer inputs. Crop yield response to organic inputs over and above response to fertilizer alone could only be assessed in three of the experiments. In one of these (Prasad and Sinha, 2000) both FYM and crop residue additions substantially increased yields of rice and wheat, with the combination of the two being the most effective (Figure 7a). At the highest fertilizer input level, the organic additions increased the yield of rice by 1.1 t/ha (35%) and the yield of wheat by 0.74 t/ha (24%). The response curves show that addition of the organic materials raised the yield plateau for both crops and that the yield increases were not simply due to increased N addition. The organic additions may have corrected an unrecognized nutrient deficiency, may have promoted an indirect effect of nutrient addition such as the effect of K on increasing resistance to lodging, or may have had some effect unrelated to nutrients such as control of soil borne pathogens. Additions of FYM had small, if any, effects on rice and wheat yields in the other two experiments (Ram, 2000; and Saha et al., 2000) (Figure 7b and c).

Results from twelve rice-wheat experiments carried out for six to ten years under the All-India Coordinated Research Project on cropping systems have been summarized by Hegde (1998a and b). The effects of supplying half of the recommended N for rice through fertilizer and half through FYM, wheat straw or *Sesbania* green manure, while maintaining recommended NPK additions to wheat, on rice and wheat yields are presented in Table 2. The yield data are the average of all years from all twelve sites, representing a total of 95 crops for both rice and wheat. Only small, and not always positive, effects of the integrated nutrient management approach were observed. Supplying half of the N through straw gave consistently lower yields than the NPK treatment. However, simply comparing treatment

FIGURE 7. Effects of organic inputs on crop yields: (a) Prasad and Sinha; (b) Ram; and (c) Saha et al.

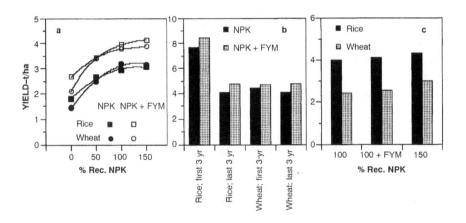

TABLE 2. Impacts of integrated nutrient management on mean crop yields for 95 crop years in All-India rice-wheat experiments.

Treatment	Mean Crop Yield[1] (t/ha)		
	Rice (R)	Wheat (W)	R + W
100% NPK	4.27	3.30	7.57
50% NPK + 50% N from FYM	3.97	3.41	7.38
50% NPK + 50% N from LGM	4.14	3.27	7.41
50% NPK + 50% N from straw	3.82	3.15	6.97

[1] Adapted from Hedge, D.M., 1998a and 1998b

means may obscure important temporal patterns in yield. Indeed, yields with organic inputs were lower than those from the NPK treatment initially but became equal to, or higher than, those achieved with NPK later on at Sabour, Bihar (Raman et al., 1996) and Faizabad, Uttar Pradesh (Alok and Yadav, 1995). Temporal data were not available from the other sites but the importance of evaluating temporal trends in order to properly assess then impact of organic inputs and/or the integrated nutrient management approach is clear.

A review by Lauren et al. (1998) of the numerous studies with LGMs in rice-wheat systems documents the substantial direct N substitution benefits of LGMs to rice and small, if any, residual effects on wheat. The authors also point out that experiments designed to determine benefits of LGMs beyond N supply, such as those listed above, are needed to better define their value.

A major constraint with almost all of the experiments with organic inputs is that neither nutrient content nor nutrient availability in the material(s) was measured. Input rates are most often based on results of generalized analyses. This approach is weak at best for crop residues and is unacceptable for FYM, which can vary considerably in quality.

Some interesting recent results that merit further study are:

- Combining different quality organic materials, typically LGMs with straw, together with inorganic fertilizer has been shown to be more effective than using the organic materials singly (Jha, Roy and Singh, 1992), possibly because of better synchrony between N supply and crop growth (Vikas, Kaur and Gupta, 1998).
- Application of "effective microorganisms" (a mixture including lactic acid bacteria, yeasts, actinomycetes and photosynthetic bacteria) together with NPK fertilizer, LGM or FYM increased crop yields and nutrient removal compared to treatment without the microorganisms (Tahir et al., 1999). Similar effects have been observed by adding cyanobacteria (Singh, Prasad and Sinha, 1997).
- Adding rice straw (5 t/ha) as a mulch to wheat in Himachal Pradesh gave yields comparable to FYM addition and higher than NPK (Verma and Sharma, 2000). Further, an increasing yield trend was found for mulch and FYM treatments, whereas yields declined with fertilizer alone. After 5 years, wheat yield in the

fertilizer plus mulch treatment was 3.6 t/ha compared to 2.4 t/ha with fertilizer alone. In contrast to the mulch treatment, straw incorporation showed the classical N immobilization depression of wheat yields in the first three years of the experiment, but yields were comparable to or higher than the no straw treatment thereafter. The straw incorporation plus FYM treatment tended to give the highest yields of both rice and wheat, suggesting that straw addition contributed to nutrient supply after the initial years where its addition caused net immobilization of N. This result contrasts with an earlier eleven year study in the Indian Punjab where straw incorporation depressed yields of both rice and wheat (Beri et al., 1995).

Effects of Organic Inputs on the Sustainability of Crop Yields

The effects of organic inputs on the sustainability of rice and wheat productivity can be assessed by comparing the average yearly yield change for treatments with and without organic additions. The paired points in Figure 8 represent the latter combinations using mean yield data. In all cases the higher yield is associated with organic additions. If productivity were more sustainable with organic additions the yearly yield change would be less negative (or more positive if yield increases were found) than without organic additions, i.e., the upper point of each pair would be to the right of the lower point. However, no consistent effect of organic additions on yield change is found for

FIGURE 8. Effects of organic inputs on sustainability of crop yields: (a) Ram; (b) Singh, Yd. et al; (c) Kundu and Samui; (d) Prasad and Sinha; and (e) Gami and Sah. Paired points represent ± organic inputs for a given experiment and the higher yield value is always + organic inputs.

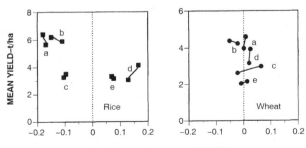

AVERAGE YEARLY YIELD CHANGE–t/ha

either rice or wheat. While organic additions increased crop productivity in these experiments, there is no compelling evidence from the few data sets available for this analysis that they generally improve the sustainability of crop yields.

Changes in Soil Organic Matter

Soil organic matter or carbon (SOC) content, which is increased by the use of organic manures has traditionally been linked with soil fertility maintenance. The organic carbon content of soils is mostly regulated via the formation and dynamics of aggregates, which protect organic matter from decomposition. The potential to form aggregates and their stability varies with soil texture and mineralogy. Aggregate formation protects organic matter by a combination of processes; via chemical complexation within micro-aggregates (< 250 μm) and via physical protection in macro-aggregates (> 250 μm). The potential to form aggregates and the reactive surface area of soil minerals increases as mineral particles become smaller, so fine textured soils generally have higher SOC levels than light textured soils. This trend is dramatically illustrated in the IGP where sandy soils generally have SOC contents in the range of 0.3 to 0.5%, while finer textured soils can have SOC contents above 1% (Table 3).

Unfortunately, puddling of soils for rice destroys soil macro-aggregates leading to substantial losses of SOC in the rice-wheat system. An example for a soil initially high in SOC is the mollisol at Pantnagar, where 20 years of cropping reduced SOC from 1.48% to 0.49% and 0.84% in unfertilized and recommended NPK treatments, respectively (Table 3; Ram). High losses of SOC in the rice-wheat system contrast with results from double and triple cropped long-term rice experiments where SOC levels are maintained or even increased (Cassman et al., 1996) due to differences in decomposition patterns and products under more continuous maintenance of anaerobic conditions.

The level of SOC is increased with inputs of organic materials to soil, including crop residues, green manures and FYM. Inputs of organic materials via roots and root exudates often increase with crop yield so that SOC levels can be higher in fertilized than unfertilized treatments (Table 3). Additions of organic materials result in small increases in SOC levels in coarse textured soils in the Punjab of India (see data from Rekhi et al., 2000 and Singh, Yd. et al., 2000) can maintain or even increase SOC levels beyond initial values in finer

TABLE 3. Effect of the rice-wheat system on soil organic carbon levels

Property/ Treatment	Ram[1]	Regmi et al.[2]	Bhattarai & Mishra	Saha et al.[4]	Rekhi et al.[5]	Yd. Singh et al.[6]	Prasad & Sinha[7]	Gami & Sah[3,8]
				Soil Organic Carbon–% C				
Initial level	1.48	1.03	–	0.71	0.18	0.36	0.57	–
Unfertilized	0.49	0.73	0.84	0.40	0.20	0.34	–	0.88
Rec. NPK	0.84	0.88	0.84	0.43	0.37	–	0.68	0.98
NPK + FYM	1.49	–	–	0.45	–	0.45	0.75	1.38
NPK + Straw	–	–	–	–	–	0.47	–	1.27
NPK + LGM	–	–	–	–	0.41	0.41	–	–
FYM only	–	1.75	1.34	–	–	–	–	1.59
Soil texture	silty clay loam	silt loam	sandy loam	sandy loam	loamy sand	loamy sand	sandy loam	sandy loam
Expt. time (years)	20	18	19	25	13	7	8	18

1 FYM added at 15 t dry wt/ha/yr before wheat.
2 FYM added at 10 t dry wt/ha/crop (30 t/ha/yr).
3 FYM added at 10 t dry wt/ha/crop (20 t/ha/yr); NPK + FYM or straw treatments are 10 t organic material plus 50 kg N/ha for rice and rec. NPK for wheat.
4 Rice-wheat-jute cropping sequence with FYM added at 10 t dry wt/ha/yr before jute.
5 LGM added at 2.3 t dry wt/ha plus 60 kg N/ha before rice with rec. NPK to wheat.
6 FYM, straw or LGM added at 6.3, 6.3, and 3.6 t dry wt/ha to rice (plus fertilizer N to bring N rate to 150 kg N) and rec. NPK to wheat.
7 FYM added at 10 t dry wt/ha prior to rice.
8 Data converted to SOC by dividing SOM values by 1.8.

textured soils. Thus addition of 15 t/ha dry weight FYM/yr maintained the original level of SOC (1.48%) at Pantanagar, India (Ram) and additions of 30 t/ha dry weight FYM/yr increased the SOC level from 1.03 to 1.78% at Bhairahawa, Nepal (Regmi et al., 2000). Unfortunately, FYM additions represent a concentration of organic residue inputs on less land than is required to generate it so it is not possible to address the widespread declines in SOC levels via manure addition.

Despite evidence that organic inputs have the expected positive effect on SOC levels, and even increased SOC above initial levels in some soils, it is clear that increasing soil organic matter *per se* does not counteract the observed yield declines in rice (Figures 2, 3 and 7).

CROP YIELD TRENDS IN NORTH WEST INDIA

Production of rice and wheat in the states of Punjab and Haryana is critical to food security in India; these states produce about 20% of the total rice and wheat production in the country and supplied 68% of the cereal grain procured by the government in 1994-5. Yield trends for wheat at the district level are shown in Figure 9 and at the state level in Figure 10a. State level data from Haryana were highly variable prior to 1973 and were excluded from the analysis. For the time periods used, wheat yields show a linear increase over time in all districts and for the two states overall. The mean annual increase ranges from 70 to 110 kg/ha and is higher in Haryana than in Punjab (Figure 10b). Although prior analysis of this data by a fact finding committee highlighted a declining rate of yield increase (expressed on % basis) over time (ICAR, 1998), yearly increases in productivity have, in fact, been constant in absolute amounts of grain. Furthermore, the data do not justify decadal analysis as was done in the earlier study. The extent to which yield increases are caused by adoption of HYVs and improved irrigation over time are not known, and increases in yield are not expected to continue indefinitely.

Rice yield trends for the states of Punjab and Haryana are curvilinear and appear to be reaching a plateau (Figure 11). District level productivity trends in Punjab (Figure 12) show some variation, with half of the districts reaching a plateau in the early 1980's. Yield trend patterns provide evidence for possible declines in productivity in Ludhiana and Sangrur districts in recent years. Yield plateaus for rice are also observed for many districts in Haryana, but data are not shown

FIGURE 9. Trends in district wheat yields for Haryana and Punjab States, India.

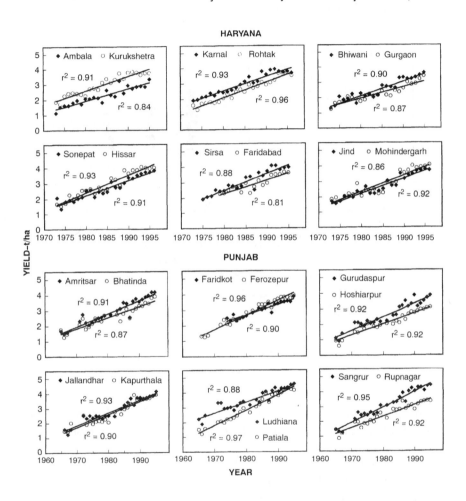

because increasing planting of lower yielding basmati varieties is a contributing factor to yield patterns in this state. Consistent with the yield trend results, increases in wheat yields have been an important contributor to gains in total production in the Punjab since the beginning of the green revolution. Figure 13 shows that yield and area planted to wheat increased proportionally between 1965 and 1983 and after that time yield has increased at a more rapid rate than area. For rice, increases in yield contributed more to total production than in-

FIGURE 10. Trends in wheat yields in (a) Haryana and Punjab States and (b) distribution of annual yield increases by district.

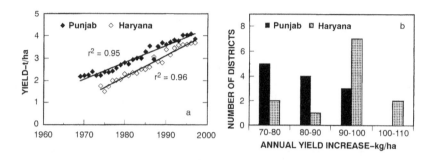

FIGURE 11. Trends in rice yields in Haryana and Punjab States.

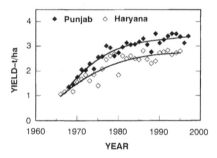

creases in area planted to rice between 1965 and 1975. However, from 1975 onwards increases in area planted to rice are largely responsible for total production gains.

The productivity trends observed in Punjab and Haryana are consistent with the general result from long-term experiments that wheat productivity is more robust than that of rice. The need for increasing inputs into the rice-wheat system, or decreasing factor productivity (Hobbs and Morris, 1996), further emphasizes that the system is under stress. Although there is a tendency to attribute declining yield trends and decreased factor productivity to declining soil nutrient supply and poor soil physical conditions for wheat, recent work with soil solarization (heating by covering with clear plastic) suggests that soil biology may be a more important constraint (Duxbury, unpublished data). This is a poorly researched area of soil science.

FIGURE 12. Trends in district rice yields for Punjab State, India.

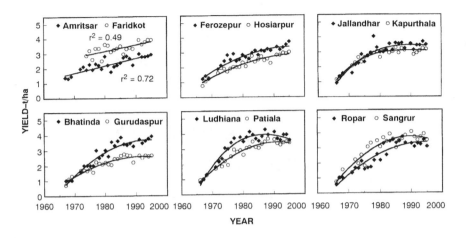

CONCLUSIONS

Rice yields in the rice-wheat system declined in most experiments, whereas wheat yields were more stable. The greater stability of wheat yields does not necessarily mean that conditions are as favorable as they could be for wheat, rather it is likely a reflection of the fact that yields are generally lower than the yield potential. Reasons for the decline in rice yields probably vary amongst experiments. The response of rice to increasing nutrient inputs was shown to drop over time in some experiments, suggesting that constraints other than N or other macronutrients contribute to yield declines. Micronutrient deficiencies and the build up of pest and pathogen pressures are possible constraints that could be evaluated through the collection of additional information.

Various organic materials proved to partially substitute for fertilizer NPK and increased soil organic matter content. However, results from the experiments did not support the widely held concepts that the use of organic materials in an integrated nutrient management approach or increases in SOC content improve the sustainability of crop yields in the rice-wheat system. In most cases, it was difficult to evaluate effects of organic additions beyond nutrient substitution because of the tendency to design experiments for this latter purpose. Interpretation of

FIGURE 13. Temporal patterns in productivity, cropped area and production for rice and wheat in Punjab State, India.

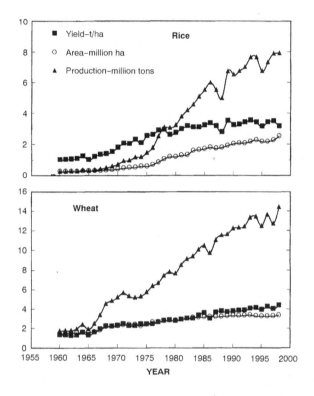

experiments is also hindered by the fact that the dry mass and nutrient content of organic inputs were not routinely determined.

There is increasing evidence that manipulating soil biological processes can give yield benefits and this area needs much more attention. Positive effects of combining high and low quality residues have been observed and appear to have the potential to improve N use efficiency by providing better synchrony between N supply and plant demand. Providing microbial supplements also appears to be a promising area for future research.

Yield trends in the northwest Indian states of Haryana and Punjab were consistent with the results from the long-term experiments with the productivity of wheat being more robust than that of rice. District level data provide strong evidence for rice yields stagnating in Punjab since the early to mid-1980's, and possibly declining yields in two districts in recent years. Whether the latter is true or not, it is clear that the rice-wheat system is a system under stress and that there are unlikely to be simple answers to its problems.

REFERENCES

Abrol, I.P., K.F., Bronson, J.M. Duxbury, and R.K. Gupta (eds.). (2000). *Long-Term Soil Fertility Experiments in Rice-Wheat Cropping Systems.* Rice-Wheat Consortium, CIMMYT, New Delhi, India.

Alok, K. and D.S. Yadav. (1995). Use of organic manure and fertilizer in rice (*Oryza sativa*)-wheat (*Triticum aestivum*) cropping system for sustainability. *Indian Journal of Agricultural Sciences* 65(10): 703-707.

Beri, V., B.S. Sidhu, G.S. Bahl, and A.K. Bhat. (1995). Nitrogen and phosphorus transformations as affected by crop residue management practices and their influence on crop yield. *Soil Use and Management* 11:51-54.

Bhattarari, E.M., and R. Mishra. (1991). Effect of long-term application of chemical fertilizer and manure on crop productivity and soil fertility under rice-wheat cropping pattern at Khajura, Nepalgunj. pp. 59-84. In Proc. 1st national workshop on long-term soil fertility experiments, Soil Science Div. NARC, Khumaltar, Nepal.

Cassman, K.G., A. Dobermann, P.C. Sta. Cruz, G.C. Gines, M.I. Samson, J.P. Descalsota, J.M. Alcantara, M.A. Dizon, and D.C. Olk. (1996). Soil organic matter and indigenous nitrogen supply of intensive irrigated rice systems in the tropics. *Plant and Soil* 32:427-444.

Gami, S.K. and M.P. Sah. (1999). Long-term soil fertility experiment under rice-wheat cropping system. pp. 12-34. In Proc. 1st national workshop on long-term soil fertility experiments, Soil Science Div. NARC, Khumaltar, Nepal.

Hegde, D.M. (1998a). Effect of integrated nutrient supply on crop productivity and soil fertility in rice (*Oryza sativa*)-wheat (*Triticum aestivum*) system in semi-arid and humid ecosystems. *Indian Journal of Agronomy* 43(1): 7-12.

Hegde, D.M. (1998b). Long-term sustainability of productivity in rice (*Oryza sativa*)-wheat (*Triticum aestivum*) system in subhumid ecosystem through integrated nutrient supply. *Indian Journal of Agronomy* 43(2): 189-198.

Hobbs, P. and M. Morris. (1996). Meeting South Asia's future food requirements from rice-wheat cropping systems: Priority issues facing researchers in the post-green revolution era. NRG paper 96-01. CIMMYT, Mexico, D.F.

Indian Council of Agricultural Research (IACR). (1998). *Decline in Crop Productivity in Haryana and Punjab: Myth or Reality?* New Delhi, India, IACR.

Jha, J.N., B. Roy, and S.P. Singh. (1992). Effect of combined-application of organic manures and inorganic fertilizer on rice and their residue on wheat. *Journal of Applied Biology* 2(1/2): 55-59.

Kundu, A.L. and R.C. Samui. (2000). Long-term soil fertility experiment in rice-wheat cropping system in West Bengal in *Long-term soil fetility experiments in rice-wheat cropping systems*, eds. I.P. Abrol, K.F. Bronson, J.M. Duxbury, and

R.K. Gupta. Rice-Wheat Consortium Paper Series 6, New Delhi, India: Rice-Wheat Consortium for the Indo-Gangetic Plains, pp. 63-67.

Lauren, J.G., J.M. Duxbury, V. Beri, M.A. Razzaque II, M.A. Sattar, S.P. Pandey, S. Bhattarai, R.A. Mann, and J.K. Ladha. (1998). Direct and residual effects from forage and green manure legumes in rice-based cropping systems. In *Residual Effects of Legumes in Rice and Wheat Cropping Systems of the Indo-Gangetic Plain*, ed. J.V.D.K. Kumar Rao, C. Johansen and T.J. Rego, ICRISAT, Pantancheru 502324, Andhra Pradesh, India. pp. 55-81.

Nambiar, K.K.M. (1994). *Soil Fertility and Crop Productivity Under Long-Term Fertilizer Use in India*. Pusa, New Delhi: Indian Council of Agricultural Research, pp. 27-28.

Nepal Agricultural Research Council (NARC), Soil Science Division. (1998). *Proceedings of First National Workshop on Long-Term Soil Fertility Experiments*, Lalitpur, Nepal, 11-13 August 1998.

Olk, D.C., K.G. Cassman, N. Mahieu, and E.W. Randall. (1998). Conserved chemical properties of young humic acid fractions in tropical lowland soil under intensive irrigated rice cropping. *European Journal of Soil Science* 49(2): 337-349.

Olk, D.C., G. Brunetti, and N. Senesi. (1999). Organic matter in double-cropped lowland rice soils: chemical and spectroscopic properties. *Soil Science* 164(9): 633-649.

Prasad, B. and S.K. Sinha. (2000). Long-term effects of fertilizers and organic manures on crop yields, nutrient balance, and soil properties in rice-wheat cropping system of Bihar. In *Long-term soil fertility experiments in rice-wheat cropping systems*, eds. I.P. Abrol, K.F. Bronson, J.M. Duxbury, and R.K. Gupta. Rice-Wheat Consortium Paper Series 6, New Delhi, India: Rice-Wheat Consortium for the Indo-Gangetic Plains, pp. 105-119.

Ram, N. (2000). Long-term effects of fertilizers on rice-wheat-cowpea productivity and soil properties in mollisols. In *Long-term soil fertility experiments in rice-wheat cropping systems*, eds. I.P. Abrol, K.F. Bronson, J.M. Duxbury, and R.K. Gupta. Rice-Wheat Consortium Paper Series 6, New Delhi, India: Rice-Wheat Consortium for the Indo-Gangetic Plains, pp. 50-55.

Raman, K.R., M.P. Singh, R.O. Singh, and U.S.P. Singh. (1996). Long-term effect of inorganic and organo-inorganic nutrient supply system of yield trends in rice-wheat cropping system. *Journal of Applied Biology* 6(1/2): 56-58.

Regmi, A.P., S.P. Pandey, and D. Joshy (2000). Effect of long-term applications of fertilizers and manure on soil fertility and crop yields in rice-rice-wheat cropping system of Nepal. In *Long-term soil fertility experiments in rice-wheat cropping systems*, eds. I.P. Abrol, K.F. Bronson, J.M. Duxbury, and R.K. Gupta. Rice-Wheat Consortium Paper Series 6, New Delhi, India: Rice-Wheat Consortium for the Indo-Gangetic Plains, pp. 120-138.

Rekhi, R.S., D.K. Benbi, and B. Singh. (2000). Effect of fertilizers and organic manures on crop yields and soil properties in rice-wheat cropping system. In *Long-term soil fertility experiments in rice-wheat cropping systems*, eds. I.P. Abrol, K.F. Bronson, J.M. Duxbury, and R.K. Gupta. Rice-Wheat Consortium Paper Series 6, New Delhi, India: Rice-Wheat Consortium for the Indo-Gangetic Plains, pp. 1-6.

Saha, M.N., A.R. Saha, B.C. Mandal, and P.K. Ray. (2000). Effects of long-term jute-rice-wheat cropping system on crop yields and soil fertility. In *Long-term soil fertility experiments in rice-wheat cropping systems*, eds. I.P. Abrol, K.F. Bronson, J.M. Duxbury, and R.K. Gupta. Rice-Wheat Consortium Paper Series 6, New Delhi, India: Rice-Wheat Consortium for the Indo-Gangetic Plains, pp. 94-104.

Sakal, R. (2000). Long-term effects of nitrogen, phosphorus, and zinc management on crop yields and micronutrient availability in rice-wheat cropping system on calcareous soils. In *Long-term soil fertility experiments in rice-wheat cropping systems*, eds. I.P. Abrol, K.F. Bronson, J.M. Duxbury, and R.K. Gupta. Rice-Wheat Consortium Paper Series 6, New Delhi, India: Rice-Wheat Consortium for the Indo-Gangetic Plains, pp. 73-82.

Singh, K.N., B. Prasad, and R.K. Sinha. (1997). Integrated effects of organic manure, biofertilizer and chemical fertilizers in rice-wheat sequence. *Journal of Research, Birsa Agricultural University* 9(1): 23-29.

Singh, Y., S.P. Singh, and A.K. Bhardwaj. (2000). Long-term effects of nitrogen, phosphorus, and potassium fertilizers on rice-wheat productivity and properties of mollisols in Himalayan foothills. In *Long-term soil fertility experiments in rice-wheat cropping systems*, eds. I.P. Abrol, K.F. Bronson, J.M. Duxbury, and R.K. Gupta. Rice-Wheat Consortium Paper Series 6, New Delhi, India: Rice-Wheat Consortium for the Indo-Gangetic Plains, pp. 14-21.

Singh, P. and R.A. Khan. (2000). Long-term effects of fertilizer practices on yield and profitability of rice-wheat cropping system. In *Long-term soil fertility experiments in rice-wheat cropping systems*, eds. I.P. Abrol, K.F. Bronson, J.M. Duxbury, and R.K. Gupta. Rice-Wheat Consortium Paper Series 6, New Delhi, India: Rice-Wheat Consortium for the Indo-Gangetic Plains, pp. 7-13.

Singh, Yd., B. Singh, O.P. Meelu, and C.S. Khind. (2000). Long-term effects of organic manuring and crop residues on the productivity and sustainability of rice-wheat cropping system of Northwest India. In *Long-term soil fertility experiments in rice-wheat cropping systems*, eds. I.P. Abrol, K.F. Bronson, J.M. Duxbury, and R.K. Gupta. Rice-Wheat Consortium Paper Series 6, New Delhi, India: Rice-Wheat Consortium for the Indo-Gangetic Plains, pp. 149-162.

Tahir, H., T. Javaid, J.F. Parr, G. Jilani, and M.A. Haq. (1999). Rice and wheat production in Pakistan with effective microorganisms. *American Journal of Alternative Agriculture* 14(1): 30-36.

Verma, T.S. and P.K. Sharma. (2000). Effects of organic residue management on productivity of the rice-wheat cropping system. In *Long-term Soil Fertility Experiments in Rice-Wheat Cropping Systems*, ed. I.P. Abrol, K.F. Bronson, J.M. Duxbury, and R.K. Gupta, Rice-Wheat Consortium, CIMMYT, New Delhi, India, pp. 163-171.

Vikas, M., B. Kaur, and S.R. Gupta. (1998). Soil microbial biomass and nitrogen mineralization in straw incorporated soils. In *Ecological Agriculture and Sustainable Development: Volume 1*. Proceedings of international conference on ecological agriculture: towards sustainable development, Chandigarh, India, 15-17 November, 1997.

Yadav, C.R., R.B. Bhujel, H.K. Prasai, and A.L. Chaudhary. (1999). Long-term fertility trial on rice-wheat-fallow cropping system at Tarahara. pp. 35-58. In Proc. 1st national workshop on long-term soil fertility experiments, Soil Science Div. NARC, Khumaltar, Nepal.

Yadav, R.L., D.S. Yadav, R.M. Singh, and A. Kumar. (1998). Long-term effects of inorganic fertilizer inputs on crop productivity in a rice-wheat cropping system. *Nutrient Cycling in Agroecosystems* 51: 193-200.

An Agroclimatological Characterization of the Indo-Gangetic Plains

J. W. White
A. Rodriguez-Aguilar

SUMMARY. The climate of the Indo-Gangetic Plains (IGP) is dominated by the Asian summer monsoon. The cool, dry winter is followed by a warming trend with daytime temperatures reaching as high as 45°C in June or July. The temperature rise is broken by the onset of the monsoon rains, when the daytime maximum temperature will immediately drop 5°C or more with the first rains. Summer temperatures are generally higher in the northwest part of the IGP, corresponding to later onset of the rainy season. In most of the IGP proper, winter temperatures is mild allowing production of wheat, potatoes and other cool season crops where irrigation is possible.

Annual precipitation varies from less than 400 mm in western Pakistan to over 1600 mm in eastern India and in Bangladesh. Based on a critical value of 0.5 for the ratio of precipitation to potential evapotranspiration, most of the IGP have abundant water for summer rainfed crops. This is reflected in the widespread cultivation of lowland rice.

Variation in rainfall over years heavily influences risk in agricultural production in the IGP. Over a 90-year period, the mean annual rainfall at Ludhiana, India was 710 mm, but in 1899, it received only 240 mm as compared to over 1400 mm in 1988. Failures of the monsoon rains cause crop failures and in the past have contributed to famines. Flooding due to excess rain can also be destructive, particularly in the eastern

J. W. White, Senior Scientist and Head, and A. Rodriguez-Aguilar are affiliated with the GIS/Modelling Laboratory, Natural Resources Group, CIMMYT Int., Lisboa 27, Apdo. Postal 6-41, 06600 Mexico, D.F., Mexico (E-mail: j.white@cgiar.org).

[Haworth co-indexing entry note]: "An Agroclimatological Characterization of the Indo-Gangetic Plains." White, J. W., and A. Rodriguez-Aguilar. Co-published simultaneously in *Journal of Crop Production* (Food Products Press, an imprint of The Haworth Press, Inc.) Vol. 3, No. 2(#6), 2001, pp. 53-65; and: *The Rice-Wheat Cropping System of South Asia: Trends, Constraints, Productivity and Policy* (ed: Palit K. Kataki) Food Products Press, an imprint of The Haworth Press. Inc., 2001, pp. 53-65. Single or multiple copies of this article are available for a fee from The Haworth Document Delivery Service [1-800-342-9678, 9:00 a.m. - 5:00 p.m. (EST). E-mail address: getinfo@haworthpressinc.com].

IGP. Attempts to predict the time of onset and amount of monsoon rains based on the El Niño-Southern Oscillation (ENSO) process and other information show promise but are not yet reliable enough for routine use in guiding agricultural management. *[Article copies available for a fee from The Haworth Document Delivery Service: 1-800-342-9678. E-mail address: <getinfo@haworthpressinc.com> Website: <http://www.HaworthPress. com>*

KEYWORDS. Climate, Indo-Gangetic Plains, monsoons, rainfall, temperature

ABBREVIATIONS. ENSO, El Niño-Southern Oscillation; IGP, Indo-Gangetic Plains; SSTA, Surface Sea Temperature Anomalies

INTRODUCTION

The Indo-Gangetic Plains (IGP) of Pakistan, India, Nepal and Bangladesh represent one of the most productive agricultural regions of the world. The relatively favorable climate is key to this productivity. Rainfall is abundant, and mild winter temperatures permit year-round intensive cropping.

The climate of the IGP, however, is by no means uniform in space or time. Precipitation and, to a lesser extent, temperature, show strong regional and year-to-year variations that have major impacts on crop production. This paper reviews the climate of the IGP with a focus on regional differences and season-to-season variation. Effects on agricultural systems are emphasized as opposed to reviewing underlying meteorological and climatic processes.

AN OVERVIEW OF THE CLIMATE

The climate of the Indo-Gangetic Plains is dominated by the Asian summer monsoon. Although summer monsoon systems exist in other parts of the world, the Asian monsoon system is by far the strongest. It is driven by heating of the large Asian landmass as the sun moves north of the equator. This creates a powerful low-pressure center that brings tropical wind and pressures systems much further north than

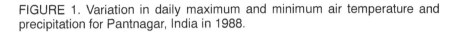

FIGURE 1. Variation in daily maximum and minimum air temperature and precipitation for Pantnagar, India in 1988.

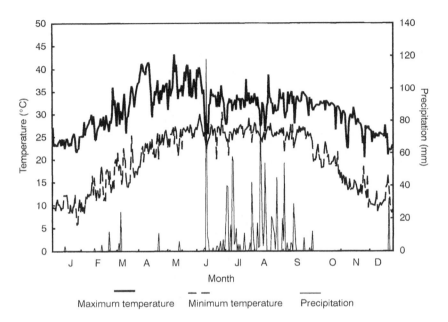

what occurs in Africa or North America. Furthermore, the band of mountains north of the IGP reduces incursions of cold air from the north.

In this system, cool, dry winter weather is followed by a warming trend with daytime temperatures reaching as high as 45° to 50°C in June or July (e.g., Figure 1 and Table 1). The trend of increasing temperature is broken by the onset of the monsoon rains which usually first reach the eastern part of the IGP. The daytime maximum temperature will drop 5°C or more with the first rains.

Summer temperatures are slightly higher in the northwest part of the IGP (Figure 2), corresponding to later onset of the rainy season. The largest temperature variations in the region, however, are associated with cooling at higher elevations, once one leaves the IGP proper.

In much of the IGP proper, winter temperatures are mild allowing production of wheat, potatoes and other cool season crops where irrigation is possible (Figure 3). However, frosts occur in Pakistan and western India.

TABLE 1. Long term monthly mean maximum and minimum temperatures and total precipitation for selected sites in the Indo-Gangetic Plains (Adapted from FAO, 1998).

	Jan.	Feb.	Mar.	Apr.	May	Jun.	Jul.	Aug.	Sept.	Oct.	Nov.	Dec.	Mean/ total
Dhaka, Bangladesh (Elev. 8 m, Lat. 23 46'N, Long. 90.23'E)													
Max. T (°C)	25.6	27.8	32.2	33.3	32.8	31.7	31.1	31.1	31.7	31.1	28.9	26.1	30.3
Min. T (°C)	12.8	15.0	20.0	23.3	24.4	25.6	26.1	26.1	26.1	23.9	20.0	13.9	21.4
Precipitation (mm)	15	21	54	108	253	420	405	324	256	153	27	3	2039
Dinajpur, Bangladesh (Elev. 37 m, Lat. 25 39'N, Long. 88 41'E)													
Max. T (°C)	24.9	27.3	32.3	35.4	33.8	32.2	31.6	31.6	31.4	31.2	28.8	25.9	30.5
Min. T (°C)	10.3	12.3	16.7	21.2	23.7	25.3	26.2	26.0	25.5	22.3	16.1	11.7	19.8
Precipitation (mm)	10	12	20	53	166	342	399	328	305	133	10	1	1779
Hissar, India (Elev. 221 m, Lat. 29.1'N, Long. 75.44'E)													
Max. T (°C)	21.7	25.0	30.7	37.0	41.6	41.3	37.3	35.5	35.7	34.6	29.6	24.1	32.8
Min. T (°C)	5.5	8.1	13.3	19.0	24.6	27.7	27.3	26.1	23.9	17.4	9.8	6.0	17.4
Precipitation (mm)	19	15	17	6	11	34	122	114	81	15	8	5	447
Ludhiana, India (Elev. 247 m, Lat. 30.56'N, Long. 75.52'E)													
Max. T (°C)	20.2	23.3	29.0	36.0	41.2	41.1	36.0	34.7	35.3	33.9	28.8	22.9	31.9
Min. T (°C)	5.8	8.4	12.9	18.5	24.2	27.1	26.7	26.1	23.9	17.5	10.1	6.2	17.3
Precipitation (mm)	32	33	27	14	15	56	205	173	113	17	3	17	705
Bhairawa, Nepal (Elev. 109 m, Lat. 27.31'N, Long. 83.27'E)													
Max. T (°C)	22.2	24.7	31.2	36.2	36.6	34.6	32.5	32.5	31.6	31.0	28.2	23.4	30.4
Min. T (°C)	6.3	9.6	13.6	19.7	23.4	24.0	25.0	25.1	23.8	20.7	13.3	8.1	17.7
Precipitation (mm)	10	0	6	7	35	214	393	367	218	61	0	0	1311
Parwanipur, Nepal (Elev. 115 m, Lat. 27.04'N, Long. 84.58'E)													
Max. T (°C)	23.2	25.6	31.9	36.2	36.7	35.0	33.7	32.8	32.1	32.1	29.1	24.9	31.1
Min. T (°C)	9.0	10.6	14.1	19.7	23.8	24.7	25.4	24.5	24.1	21.6	14.1	9.0	18.4
Precipitation (mm)	12	9	19	32	54	203	324	294	209	55	8	5	1224
Dera Ismail Khan, Pakistan (Elev. 172 m, Lat. 31.49'N, Long. 70.55'E)													
Max. T (°C)	20.0	22.0	27.7	33.7	39.7	42.1	39.6	38.1	37.4	34.1	27.7	21.9	32.0
Min. T (°C)	4.6	7.2	12.8	18.4	23.7	27.5	28.2	27.3	24.2	16.5	9.3	5.1	17.1
Precipitation (mm)	14	18	27	20	9	9	65	36	14	2	3	6	223

	Jan.	Feb.	Mar.	Apr.	May	Jun.	Jul.	Aug.	Sept.	Oct.	Nov.	Dec.	Mean/total
Islamabad, Pakistan (Elev. 508 m, Lat. 33.37′N, Long. 73.06′E)													
Max. T (°C)	16.3	20.1	23.6	29.7	35.5	39.7	35.5	33.2	33.5	29.8	24.2	19.1	28.4
Min. T (°C)	2.7	4.8	10.3	14.6	20.2	24.4	24.9	23.8	21.8	14.4	7.2	3.4	14.4
Precipitation (mm)	47	50	44	33	27	45	165	190	71	11	6	23	712
Lahore, Pakistan (Elev. 214 m, Lat. 31.33′N, Long. 74.2 120′E)													
Max. T (°C)	19.3	22.4	27.8	34.8	40.2	41.1	37.1	35.8	35.9	33.7	27.9	21.9	31.5
Min. T (°C)	5.1	8.1	13.1	18.4	23.1	26.8	27.2	26.6	24.1	17.3	9.6	5.7	17.1
Precipitation (mm)	24	22	23	13	15	38	150	131	63	8	2	13	502

FIGURE 2. Variation over the IGP for mean maximum temperature for the three hottest months. Based on interpolated data for mean monthly temperatures from P. Jones (1999, personal communication) with additional processing using the SCT (Corbett and O'Brien, 1997).

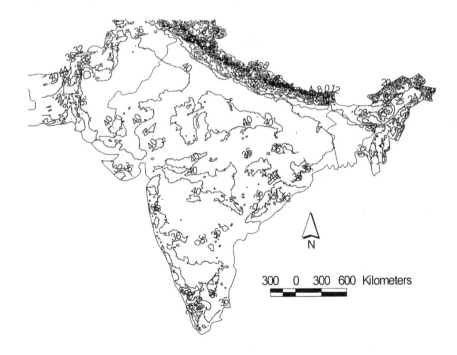

FIGURE 3. Variation over the IGP for mean minimum temperature for the three coolest months. Data source is as in Figure 2.

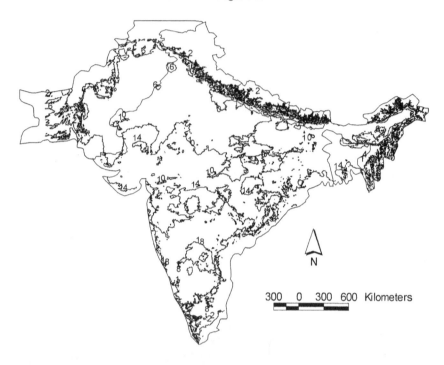

Annual precipitation within the IGP varies from less than 400 mm in parts of Pakistan to over 1600 mm in eastern India and in Bangladesh (Figure 4). Neighboring hill regions can receive much larger amounts, with locations such as Lumle, Nepal reporting an annual totals over 4000 mm.

A critical value of 0.5 for the ratio of precipitation to potential evapotranspiration (P/PE) is often used to indicate whether rainfed agricultural production is feasible. Most of India, Nepal and Bangladesh have P/PE values well in excess of this limit (Figure 5).

SIMILARITIES TO CLIMATES ELSEWHERE IN THE WORLD

The climate of the IGP shows limited similarities with regions with summer monsoon rains, reflecting the uniqueness of the Asia mon-

FIGURE 4. Variation in total annual precipitation (mm) over the IGP. Data source is as in Figure 2.

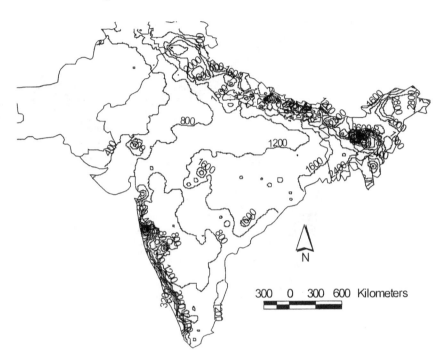

soon system. However, if one focuses on growing seasons for specific crops, important similarities are seen. One example is the west-central coast of Mexico where the Green Revolution wheat technologies were first developed. The winter climate of this region shows similarities to large areas of the IGP (Figure 6).

CLIMATIC RISK

Variation in rainfall over years is a key determinant of food security in the IGP. Ludhiana, India has mean annual rainfall of 710 mm but in 1899, it received only 240 mm as compared to over 1400 mm in 1988. Failures of the monsoon rains have caused crop failures and contributed to famines.

However, it is important to understand that a drought in one location of the IGP does not always imply low rainfall throughout the

FIGURE 5. Variation over the IGP for the ratio of precipitation to potential evapotranspiration for the three wettest months. Data source is as in Figure 2.

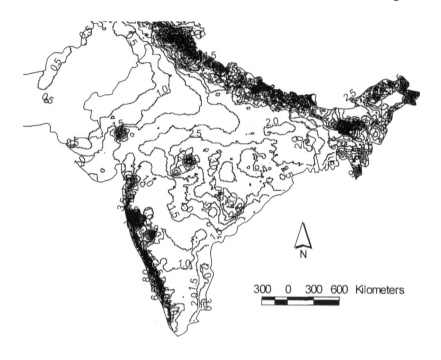

region. Correlations among annual rainfall totals at different sites are low (Table 2), and for agricultural planning, it may be more appropriate to identify sub-regions with similar rainfall regimes (Gadgil, 1989).

Flooding due to excess rain in the IGP and surrounding mountains can be destructive, particularly in the eastern IGP. However, flooding can also be viewed as a normal feature of the ecosystem, which restores soil fertility and permits activities such as fish farming in temporary ponds.

Another aspect of climatic risk in the IGP that affects agriculture is rainfall outside of the normal monsoon period. Excessive soil moisture can delay sowing of wheat and other crops after the summer rice crop (Fujisaka, Harrington and Hobbs, 1994). Rains exacerbate this problem in October and November. While long-term mean rainfalls are low during these months (Table 1), mean values hide years with rainfall totals that are considerably higher (Figure 7). This uncertainty is a

FIGURE 6. Distribution of regions within the IGP with a winter climate (coolest three months) similar to Ciudad Obregon, Mexico where Green Revolution wheats were first selected. The criteria for similarity were all regions having rainfall within ±100 mm and maximum and minimum temperatures within ± 2°C of Obregon for the coolest consecutive three months. Source of base data is as in Figure 2.

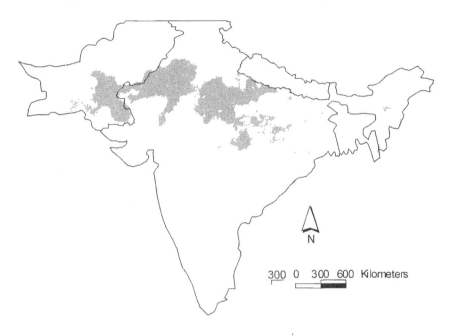

major impetus for seeking alternate methods to establish crops after rice by using practices such as surface seeding and reduced or zero tillage (Hobbs, Giri and Grace, 1997).

CLIMATE PREDICTION

Given the importance of the monsoon cycle, it is not surprising that meteorologists have long sought to predict the strength of the coming rainy season. As early as 1880, H. Blanford, the first Imperial Meteorological Reporter to the Government of India noted that the monsoon failure of 1878 was linked to very high atmospheric pressures over much of Asia. In 1885, the Indian Meteorological Department published summer monsoon rainfall forecasts (Allan, Lindesay and Parker, 1996).

TABLE 2. Correlations among annual rainfall totals for Peshawar, Pakistan and Ludhiana, Agra, Allahabad, Patna and Calcutta, India, representing a series running west to east across the IGP. Data are for 93 years from 1868 to 1960 from the Global Historical Climate Network (adapted from Vose et al., 1992).

	Mean annual rainfall (mm)	Peshawar	Ludhiana	Agra	Allahabad	Patna
Peshawar	340					
Ludhiana	710	0.20				
Agra	710	−0.03	0.29**			
Allahabad	1000	0.00	−0.01	0.33**		
Patna	1150	0.02	−0.04	0.02	0.24*	
Calcutta	1600	0.05	−0.06	−0.04	−0.04	0.18

*, ** Correlations significant at the P = 0.05 and P = 0.01 levels, respectively.

FIGURE 7. Frequency of total rainfall in November (as percentages) falling in three ranges (0 to 5 mm, 5 to 25 mm and over 25 mm) for Lahore, Pakistan and Ludhiana and Patna, India.

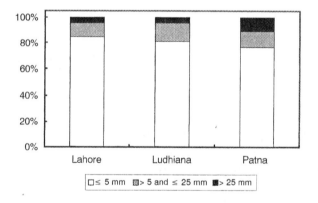

There is a strong relation between the monsoon rains and various indices based on the El Niño-Southern Oscillation (ENSO) process and other information. As a single example, Figure 8 presents the relation between the all-India monsoon rainfall index, which averages rainfall for 29 geographical subdivisions (Parthasarathy, Rupa Kumar and Munot, 1993) and June to September Surface Sea Temperature Anomalies (SSTA; Kaplan et al., 1998). Unfortunately, this relation weakens as one moves to SSTA values for month's preceding the

FIGURE 8. Relation between the All-India monsoon rainfall index (Parthasarathy et al., 1993) and June to September SSTA (Kaplan et al., 1998) from 1871 to 1994.

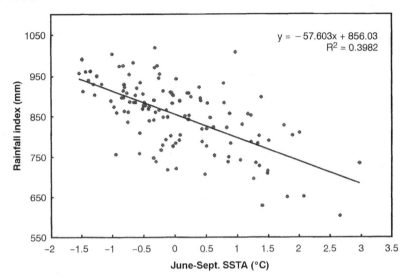

monsoon season, and used alone, SSTA values are of limited use for predicting monsoon conditions.

Various groups are using more sophisticated models for seasonal forecasts of rainfall and temperatures, typically with ranges of one to three months in advance (e.g., Shukla and Mooley, 1987; Webster et al., 1998; European Centre for Medium-Range Weather Forecasts, 1999). These approaches show promise and merit closer attention by agronomists.

A COMMENT ON DATA SOURCES

Two recent improvements in data availability greatly facilitate climatic analyses in relation to agricultural problems of the IGP. This is reflected in the data presented herein and merits further commentary.

The Internet now provides access to many climatic data sets that can be freely down-loaded for non-commercial purposes. These include monthly and annual temperature records dating to the 1860's, rainfall indices, various indices related to the ENSO, and sets of daily data weather from the 1970's onward (Table 3).

Although not Internet accessible due to the size of the data sets,

TABLE 3. Some useful sources of climate data for the IGP that are accessible through the Internet.

Data description	Internet address (URL)	Other references
All-India monsoon rainfall index from 1871 to 1993	http://grads.iges.org/india/allindia.html	Parthasarathy et al., 1993
Reconstructed sea surface temperatures from 1856 to 1991	http://ingrid.ldeo.columbia.edu/SOURCES/.KAP LAN/.EXTENDED/.ssta/	Kaplan et al., 1998
Daily temperatures, rainfall and other variables from 1994 onward	http://www.ncdc.noaa.gov/ol/climate/climatedata.html#DAILY	
Monthly and annual precipitation and temperatures from 1860s onward	http://www.ncdc.noaa.gov/ghcn/ghcn.SELECT.html	Vose et al., 1992
Long term monthly means for temperature, precipitation, potential evapotranspiration, and solar radiation	http://www.fao.org/waicent/faoinfo/agricult/agl/aglw/climwat.html	

interpolated climate data such as those used to produce Figures 2 through 6 are also proving invaluable. Interpolating long-term climate data from large numbers of locations produces the surfaces. The statistical procedures typically include effects of elevation. The monthly surfaces used in this paper were developed by Peter Jones (personal communication, 1999) at the International Center for Tropical Agriculture (CIAT, Cali, Colombia) using a modified inverse distance method. Hutchinson (1995) has developed an alternative procedure using thin plate smoothed splines that is also very effective (Hartkamp et al., 1999). Subsequent processing of the monthly surfaces was done with a geographic information system (GIS). The specific climate models (e.g., coolest quarter used in Figure 3) and the site similarity analysis (Figure 6) were produced using the Spatial Characterization Tool developed by Corbett and O'Brien (1997).

REFERENCES

Allan, R., J. Lindesay and D. Parker. (1996). *El Niño Southern Oscillation and Climatic Variability*. CSIRO; Colinwood, Victoria, Australia.

Corbett, J.D. and R.F. O'Brien. (1997). *The Spatial Characterization Tool. Texas A&M University System, Blackland Research Center Report N. 97-03*, CDROM

Publication. Texas A&M University System, Blackland Research Center; Temple, Texas.

FAO. (1998). *CLIMWAT, a climatic database for CROPWAT.* [Online]. Available at http://www.fao.org/WAICENT/FAOINFO/AGRICULT/AGL/aglw/climwat.htm (posted 4 February, 1998; verified 2 Dec., 1999).

European Centre for Medium-Range Weather Forecasts. (1999). ECMWF Seasonal Forecasting (http://.ecmwf.int/services/seasonal/forecast/index.jhtml).

Fujisaka, S., L. Harrington and P. Hobbs. (1994). Rice-wheat in South Asia: systems and long-term priorities established through diagnostic research. *Agricultural Systems* 46: 169-187.

Gadgil, S. (1989). Monsoon variability and its relationship to agricultural strategies. p. 249-256. In *Climate and Food Security.* Proceedings of the International Symposium on Climate Variability and Food Security in Developing Countries, 5-9 February, 1987, New Delhi, India. International Rice Research Institute; Manila, Philippines.

Hartkamp, A.D., K. De Beurs, A. Stein and J.W. White. (1999). *Interpolation Techniques for Climate Variables.* NRG-GIS Series 99-01. CIMMYT; Mexico, D.F., Mexico.

Hobbs, P. R., G.S. Giri and P. Grace. (1997). *Reduced and Zero Tillage Options for the Establishment of Wheat After Rice in South Asia.* RCW Paper No. 2. Rice-Wheat Consortium for the Indo-Gangetic Plains and CIMMYT; Mexico, D.F.

Hutchinson, M.F. (1995). Interpolating mean rainfall using thin plate smoothing splines. *International Journal of Geographic Information Systems* 106: 211-232.

Kaplan, A., M. Cane, Y. Kushnir, A. Clement, M. Blumenthal, and B. Rajagopalan. (1998). Analyses of global sea surface temperature 1856-1991, *Journal of Geophysical Research* 103: 18,567-18,589.

Parthasarathy, B., K. Rupa Kumar, and A.A. Munot. (1993). Homogeneous Indian Monsoon rainfall: Variability and prediction. *Proceedings of the Indian Academy of Sciences* 102: 121-155.

Shukla, J., and D.A. Mooley. (1987). Emperical prediction of the summer monsoon rainfall over India. *Monthly Weather Review,* 115, 695-703.

Vose, R.S., Richard L. Schmoyer, Peter M. Steurer, Thomas C. Peterson, Richard Heim, Thomas R. Karl, and J. Eischeid. (1992). *The Global Historical Climatology Network Long-Term Monthly Temperature, Precipitation, Sea Level Pressure, and Station Data.* ORNL/CDIAC-53, NDP-041. Carbon Dioxide Information Analysis Center, Oak Ridge National Laboratory, Oak Ridge, Tennessee.

Webster, P.J., V.O. Magana, T.N. Palmer, J. Shukla, R.A. Tomas, M. Yanai, and T. Yasunari. (1998). Monsoons: Processes, predictability, and the prospects for prediction. *Journal of Geophysical Research* 103(C7): 14,451-14,510.

Legumes and Diversification of the Rice-Wheat Cropping System

J. G. Lauren
R. Shrestha
M. A. Sattar
R. L. Yadav

SUMMARY. Grain, forage and green manure legumes have a long heritage in South Asia. This paper attempts to summarize the merits of legumes from historical, cultural, human nutrition, animal nutrition and agronomic perspectives. Despite well established and numerous benefits associated with legumes in South Asia, their production are static or declining in the region. Niches for including legumes within the rice-wheat cropping system are described. Constraints have been identified, and promising options for higher productivity suggested. *[Article copies available for a fee from The Haworth Document Delivery Service: 1-800-342-9678. E-mail address: <getinfo@haworthpressinc.com> Website: <http://www.HaworthPress.com> © 2001 by The Haworth Press, Inc. All rights reserved.]*

KEYWORDS. Grain legumes, forage legumes, green manures, human nutrition, animal nutrition, biotic constraints, crop improvement, and government policies

J. G. Lauren is affiliated with the Department of Soils and Crops, Cornell University, Ithaca, New York, USA. R. Shrestha is Agronomist, Nepal Agricultural Research Council, Kathmandu, Nepal. M. A. Sattar is affiliated with the Bangladesh Institute of Nuclear Agriculture, Mymensingh, Bangladesh. R. L. Yadav is Director, Project Directorate for Cropping Systems Research, Modipuram, India.

Address Correspondence to: J. Lauren, 917 Bradfield Hall, Cornell University, Ithaca, NY 14853 USA (E-mail: JGL5@cornell.edu).

[Haworth co-indexing entry note]: "Legumes and Diversification of the Rice-Wheat Cropping System." Lauren, J. G. et al. Co-published simultaneously in *Journal of Crop Production* (Food Products Press, an imprint of The Haworth Press, Inc.) Vol. 3, No. 2(#6), 2001, pp. 67-102; and: *The Rice-Wheat Cropping System of South Asia: Trends, Constraints, Productivity and Policy* (ed: Palit K. Kataki) Food Products Press, an imprint of The Haworth Press, Inc., 2001, pp. 67-102. Single or multiple copies of this article are available for a fee from The Haworth Document Delivery Service [1-800-342-9678, 9:00 a.m. - 5:00 p.m. (EST). E-mail address: getinfo@haworthpressinc.com].

HISTORICAL AND CULTURAL PERSPECTIVE

Humankind has long recognized the value of legumes for food, fodder and soil fertility. Domestication and cultivation of some legume species have been recorded as far back as 8000 BC along with barley and wheat (Zohary and Hopf 1973). Several grain legumes are thought to originate in South Asia, including chickpea/Bengal gram (*Cicer arietinum*), pigeonpea (*Cajanus cajan*), greengram/mungbean (*Vigna radiata*), blackgram (*Vigna mungo*) and cowpea (*Vigna unguiculata*). These pulse crops were adopted early in South Asia (circa 1000-2000 BC), thereby supplying religious vegetarians and the poor with inexpensive protein sources (Baldev 1988). While cultivation of legumes specifically for fodder is a relatively recent practice, feeding legume residues to livestock is a very old and common custom. In addition prior to the availability of chemical fertilizers, South Asians regularly cultivated legume species for green manuring as long ago as 1000 BC (Raychaudhuri 1960).

As a result of this long history in South Asia, a multitude of food preparations with grain legumes from simple dal to dosas or kabuli channa to sweets such as laddoo became an integral part of the daily life and culture of the region. Besides day-to-day uses, grain legumes play important roles in religious and cultural events. For example greengram, chickpea and pigeonpea are referred to in the oldest sacred writings of Hinduism, the Vedas, with praise and reverence. A hymn in the Vedas states "O, Agni (fire), who consumest flesh, the black goat is your share, lead is said to be your wealth, and ground blackgram said to be your offering" (quoted from Baldev 1988-*Atharva Veda*, *Anuvaka* 474, hymn 63). In Nepal, roasted black-seeded soybean and flattened rice is a common offering to a Newari goddess. Likewise in the fall months of August/September, a traditional legume soup "kwati" is prepared by mixing more than nine different grain legumes together on the occasion of Jamnai Purne (the changing of a sacred thread which Brahmin and Chetri caste Hindu males and Newar priests wear around the torso). And in northwest India, marriages and special festivals are often marked with mungbean halwa sweets (Singh, Chhabra, and Kharb 1988).

BENEFITS FROM LEGUMES

Human Nutrition

Grain legumes are well known for their high protein concentrations, ranging from 17 to 40% (Bressani and Elias 1978). Protein contents in paddy rice or wheat are substantially lower; however, when production is considered, wheat and rice are often similar or higher in protein yield than the pulses (Table 1). Perhaps more important are the differences in protein quality that exist between cereals and pulses. Wheat is deficient in the amino acid, lysine, while rice has inadequate levels of lysine and threonine (Deshpande, Singh, and Singh 1991). By contrast, grain legumes are deficient in sulfur containing amino acids, methionine and cystine (Jansen 1974). Combined consumption of cereals and grain legumes is a well known practice for overcoming the respective amino acid deficiencies in cereals and pulses, thereby achieving near complete protein balance and nutritional improvement in cereal based diets. Studies of the protein complementarity between legumes and cereals show that maximum protein is obtained when the grain legume content is around 10% in a wheat-pulse diet, while with rice, maize or barley, maximum protein is found when the pulse content is around 20% of the diet (Chatterjee and Abrol 1975).

Often overlooked are the significant concentrations of minerals and vitamins also found in grain legumes, such as folic acid, the B vitamins, riboflavin, thiamin and niacin as well as calcium, zinc and iron (Table 2). Iron, zinc, thiamin and niacin contents in wheat are on par

TABLE 1. Comparison of protein yields between cereals and grain legumes.

Crop	Average Yields in South Asia (t ha^{-1})	Protein Content (%)	Average Protein Yield (kg ha^{-1})
Cereals			
Paddy Rice	2.78	7.5	208.5
Wheat	2.17	12.0	260.4
Grain Legumes			
Soybean	1.01	38.0	383.8
Lentil	0.73	29.6	216.1
Beans	0.57	22.0	125.4
Chickpea	0.74	20.0	148.0

Adapted from Norman et al. 1995; FAOSTAT 1999; Bressani and Elias 1974.

TABLE 2. Vitamin and mineral content of select grain legumes compared with rice and wheat.

Crop	Calcium[1]	Iron[2]	Zinc[2]	Folic Acid[1]	Thiamin[3]	Niacin[3]	Riboflavin
	(mg/100 g)			(µg/100 g)	(mg/100 g)		
Chickpea	105	8.3	2.2	557	0.46	1.2	0.20
Lentil	51	10.1	2.8	433	0.46	1.3	0.33
Mungbean	132	6.9	2.3	625	0.53	2.5	0.26
Pigeonpea	130	6.8	2.9	456	0.49	2.2	0.21
Soybean	277	11.5	3.9	375	0.66	2.2	0.22
Rough Rice	40	4.2	1.8	20	0.40	5.1	0.09
Wheat	50	7.8	2.6	41	0.42	4.9	0.11

[1]USDA 1999, [2]Kumar and Kapoor 1984, [3]Pulses: Aykroyd et al. 1982; Cereals: USDA 1999.

with legumes, but calcium, folic acid and riboflavin are clearly lower than legumes. Relative to pulses and wheat, paddy rice is lower in all the considered minerals and vitamins, which is aggravated by the rice milling process. For example, thiamin levels drop from 0.26-0.33 mg/100 g to 0.02-0.11 mg/100 g with milling, and iron contents decrease from 1.4-6.0 mg/100 g to 0.2-2.8 mg/100 g with milling (FAO 1993).

Although significant gains have been made towards reducing protein-energy malnutrition through dramatic increases in cereal production, consumption of critical vitamins and minerals in grain legumes is decreasing in South Asian diets. The current health status of the populace in these countries is a direct reflection of the low supply. For example, a 1996 nationwide study by the Indian National Nutrition Monitoring Board found that the diets of children and adult women only fulfilled 47-59% of recommended riboflavin requirements (Bamji and Lakshmi 1998). Likewise folic acid consumption amongst Indian children and women were only 33-55% of recommended levels in 1976, and most likely have not improved since then (Babu 1976; Bamji and Lakshmi 1998). Regional estimates of iron deficiency anemia in young children and women of child-bearing age are 47-77%, 30-95%, 60-78% and 45% in Bangladesh, India, Nepal and Pakistan, respectively (FAO, 1998a; 1998b; 1999; Seshadri 1997). Finally stunting in children under 5 years old are currently estimated at 55%, 20-83%, 14-65% and 50% in Bangladesh, India, Nepal and Pakistan, respectively (FAO 1998a; 1998b; 1999; Vaidya 1990); while wasting measures are at 18%, 8-32%, 7-12% and 6-10% in Bangladesh, India, Nepal and Pakistan, respectively (FAO 1998a; 1998b; 1999; Vaidya

1990). Historically child wasting and stunting was associated with protein-energy malnutrition; but lately these anthropometric measures also have been related to nutrient deficiencies of zinc, iron, calcium and folate (Golden and Golden 1981; Chwang, Soemantri, and Pollitt 1988; Gopalan 1998).

Antinutritional factors found in food legumes can reduce bioavailability or cause toxic reactions (Deshpande and Deshpande 1991). Phytates are of particular concern since they complex proteins and minerals such as calcium, zinc or iron, making them unavailable for human nutritive benefit. Fortunately various cooking and processing techniques can substantially remove or denature phytates and many other antinutritional compounds. Chang, Schwimmer, and Burr (1977); Deshpande, Sathe, and Salunkhe (1984) showed reductions in phaseolus bean and black gram phytates of 50-80% through soaking and cooking procedures.

Some legume grains have the additional advantage of reducing cholesterol and blood sugar in humans. Dietary fiber from chickpea and mungbean has been reported to reduce the level of cholesterol (Jaya, Venkataraman, and Krishnamurthy 1979; Thomas, Soni, and Singh 1987), while chickpea dal and kidney bean has reduced plasma glucose levels in humans (Dilawari et al. 1981).

Animal Nutrition

Forage cultivation has never been very extensive in South Asia, even though there is a significant deficit of fodder material in the region. Traditionally farmers have invested very little capital in animal feed, but supplying tree loppings, weeds and crop residues, rather than high quality forages.

Livestock diets dominated by cereal straws are generally low in palatability and digestibility, and deficient in protein and essential minerals (Schiere, Ibrahim, and Rond 1985). It is noteworthy that India's recent impressive milk production increases were achieved by the traditional approach without major increases in quality forage production (Kurien 2000). Nevertheless protein concentrates from cereal by-products, oil seed cakes and meals have had to partially correct the nutritional deficits in the current system, thereby increasing the costs of production (Patil et al. 1987). Recent annual imports of feed concentrates to India have increased up to 53,000 Mt (FAOSTAT 1999).

Poor quality livestock feeds can also be improved by supplementation with legume forage materials. Cereal crop residues supplemented with forage legumes significantly increase liveweight gains and overall animal productivity (Moog 1986; Peoples and Herridge 1990). For example cowpea forage fed in combination with maize fodder increased liveweight gain in buffalo calves by 10% over maize fodder alone (Ahuja et al. 1991), and milk production of 4.5 L day^{-1} was sustained, which gained the farmer a net profit of Rs 5600 ha^{-1} (Rekib 1980). The benefits from legume supplements are not limited to cows or buffalo; a study from Bangladesh reported improved poultry liveweight gains of 6% and 53% greater egg production when pulse grains were included in feed (Rukonuddin, Kabirullah, and Khan 1994).

Besides benefiting the health and production of livestock, consumption of forage legumes can help reduce of methane (CH_4) greenhouse gas emissions. Ruminant animals (cattle, buffalo, sheep, and goats) produce about 65-85 $\times$ 10^{12} g of CH_4 annually, adding to the increasing atmospheric concentrations of 10 ppbv y^{-1} (Mosier et al. 1998). The work by Leng (1991) indicates that improved digestion of poor quality feeds through strategic supplementation with feed concentrates, forage legumes or chemically treated straw could reduce CH_4 by 30 to 50%. In terms of milk production, these reduction strategies could decrease CH_4 from 250 g CH_4 L^{-1} to 40-80 g CH_4 L^{-1} milk.

Agronomy

Agronomic benefits to rice or wheat from grain, forage or green manure legumes are well documented (Ahlawat et al. 1998; Buresh and DeDatta 1991; Lauren et al. 1998; Saraf et al. 1998). Much research has focussed on N benefits, because the ability of leguminous plants to biologically fix atmospheric nitrogen provides a relatively low cost method for replacing nitrogen removed through agriculture and for building soil N pools. Also there has been great interest in measuring N_2 fixation rates to determine the limitations on legume productivity or biological nitrogen fixing (BNF) capacity (Ladha, Watanabe, and Saono 1988; Peoples and Herridge 1990).

There is little doubt that legumes contribute to significantly higher yields in succeeding rice or wheat crops. Direct effects are well summarized in review articles by Peoples and Craswell (1992); Yadvinder-Singh, Khind, and Singh (1991); and Becker, Ladha, and Ali (1995).

Ahlawat et al. (1998) reported that rice yields after early summer grain legumes were 0.76 to 1.77 t ha^{-1} higher than rice following fallow or a maize fodder. Wheat yields after summer soybean, pigeonpea, or mungbean were 1.3 to 1.6 t ha^{-1} greater than wheat following pearl millet or sorghum. Similarly substituting winter grain legumes, such as lentil, chickpea, or pea for wheat increased subsequent rice yields from 190 to 550 kg ha^{-1} (Saraf et al. 1998). Reported yield responses from soils amended with green manure legumes ranged from 0.4 to 4.1 t/ha relative to controls without green manure (Lauren et al. 1998). Rahman (1994) reported that incorporating berseem (*Trifolium alexandrinum*), a forage crop, between *aman* (summer) rice and *boro* (winter) rice increased the grain yield of the succeeding *boro* rice by 7-18% and straw yield by 35%. Carangal et al. (1994) and Ladha et al. (1996) also noted rice yield increases of 0.6 to 2.4 t ha^{-1} following forage legumes.

Residual effects in a second crop after the legume are much less dramatic than with the first crop. In India Boparai et al. (1992); Rekhi and Bajwa (1993); Kolar, Grewal, and Singh (1993); and Gill, Singh, and Rana (1994) found quite low responses in wheat of only 4-11% over controls. In Bangladesh residual fertility from a berseem forage legume cropped before a *boro* rice crop increased the grain yield of the second rice crop by 4-14% and straw yield by 2-9% (Rahman 1994). Rice biomass in Nepal increased by 3-t ha^{-1} with every 5-t ha^{-1} of legume residual dry matter incorporated followed by a 17-81% increase in spring potato yields (Khatri and Wells 1999). The bulk of studies reported intermediate wheat yield increases by 15 to 38% (Sharma, Prasad, and Singh 1995; Sharma and Mittra 1988; Bhardwaj, Prasad, and Singh, 1981; Chaudhary 1990; Rathore, Chipde, and Pal 1995). Tiwari, Tiwari, and Pathak (1980a); Tiwari, Pathak, and Ram (1980b); Mahapatra, Sharma and Sharma (1987); Mahapatra and Sharma (1989); and Mann and Ashraf (1997) measured relatively high residual effects between 54 and 94%. In addition legumes contributed to significant increases in wheat N uptake ranging from 4-17 kg/ha (Goswami et al. 1988; Rekhi and Bajwa 1993; Sharma, Prasad, and Singh 1995).

Research records give the impression that N is the only contributing factor to observed yield responses in rice or wheat from legumes. However, several workers have established clearly that legumes can have an effect on the overall yield potential that no amount of fertilizer

N can overcome (Dargan, Chillar, and Bhardwaj 1975; Chatterjee et al. 1979; Tiwari, Tiwari, and Pathak 1980a; Rekhi and Meelu 1983; Beri, Meelu, and Khind 1989). Improvements in soil physical properties (Yaacob and Blair 1981; De Datta and Hundal 1984; Boparai, Yadvinder-Singh, and Sharma 1992); increased acquisition and mobility of macronutrients (Yadvinder-Singh et al. 1988; Nagarajah, Neue, and Alberto 1989) and micronutrients (Thind and Chahal 1983; BRRI 1985; Benbi and Brar 1992); as well as reductions in soil-borne pests and pathogens (Reddy et al. 1986; Pariselle and Rinaudo 1988) have all been demonstrated in cropping systems where legumes have been included.

PRODUCTION, AVAILABILITY AND ACCESSIBILITY

Despite the well established and numerous benefits associated with legumes in South Asia, production is static or declining in the region. Over the past 35 years, pulse production has stayed more or less the same in India and Pakistan, while small increases in pulse production have been noted in Nepal and Bangladesh (Figure 1). Nevertheless these increases have not kept pace with population as seen in per capita pulse availability trends (Figure 2). India and Pakistan exhibit declines in pulse supply at rates of approximately 200 g per person per year, while pulse availability in Nepal and Bangladesh has remained constant at 5 and 8 kg per person per year, respectively. Such low availability provides only 3-7% of daily caloric needs and 4-12% of protein requirements (FAO and WHO, 1973; FAO, 1998a; 1998b; 1999).

After the Green Revolution, the more productive and less risky cereals, rice and wheat (Yadav et al., 1998) gradually replaced grain legumes. As a result, pulse prices have increased to varying extents following supply and demand (Figure 3). Efforts to correct deficits in supply and demand for pulses have resulted in increased imports. Since 1987 the proportion of pulses imported into India to meet food demands has averaged about 5% of the total supply, whereas pulse imports have been as high as 26% (1990) in Bangladesh, 18% in Nepal and 19% in Pakistan (1994) (FAOSTAT, 1999). Such requirements are difficult burdens for these countries given their low foreign exchange capacities.

FIGURE 1. Pulse production in South Asia, 1961-1998 (FAOSTAT 1999).

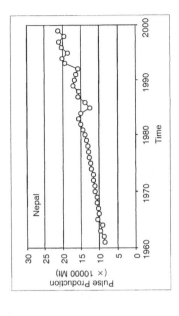

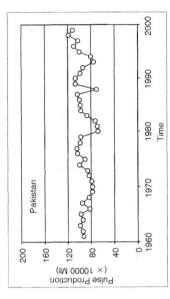

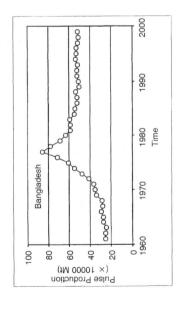

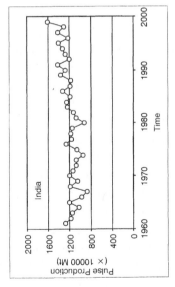

FIGURE 2. Per capita pulse supplies in South Asia, 1961-1997 (FAOSTAT 1999).

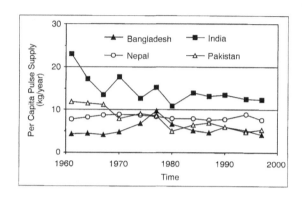

Devendra (1997) estimated that annual livestock feed availability from cereal residues alone is 1.2, 0.9, 1.7 and 0.6 tons per ruminant animal in Bangladesh, India, Nepal and Pakistan, respectively. Based on a "good" average annual intake of 1.28 t for buffalo and 0.73 t for cattle, one can see that substantial gaps between ruminant animal feed demand and supply exist for India and Pakistan. Such an analysis excludes feed needs for non-ruminant animals (goats, sheep, and poultry) and recognizes that quality forage supplements are necessary for good livestock productivity. In India alone, it is estimated that 1.1 billion tons of green forage would be required for livestock populations by the year 2000 (Singh 1988). Likewise in Pakistan, total digestible nutrients and digestible crude protein are in deficit by 57% and 39%, respectively (Singh et al. 1997). Despite these clear deficits, forage legume cultivation in South Asia is still very limited, primarily because of land competition with intensive cereal production. For example, only 4% of India's cultivated area (6.6 million ha) is devoted to forage production (Singh et al. 1997).

Cultivation of green manure legume species has declined since the Green Revolution. Although little data is available about current farmer practice, it appears that <5% of the area in Asia as a whole (6 million ha) receives additions of legumes as green manures (Garrity and Flinn 1988; Lizhi 1988).

FIGURE 3. Trends in South Asia pulse prices, 1965-1994 (FAOSTAT 1999).

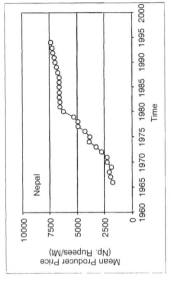

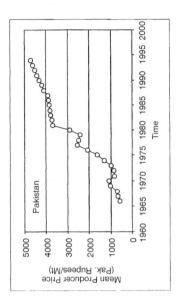

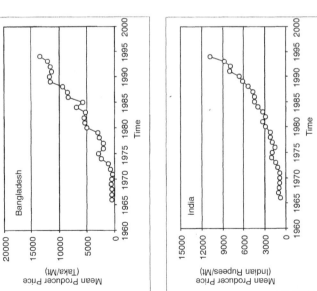

LEGUME OPTIONS FOR RICE-WHEAT

Grain Legumes

Area and productivity for the dominant grain legume species culti-vated in Bangladesh, India, Nepal and Pakistan are presented in Table 3. Lathyrus, lentil, chickpea, black gram, mungbean, pigeonpea and soy-bean are often included within rice-wheat rotations of South Asia as substitutes for rice or wheat or as relay/catch crops between the cere-als. Winter legumes lathyrus, lentil and chickpea substitute for wheat or are intercropped with wheat followed by summer rice (Miah and Rahman 1993; Neupane and Bharati 1993; Yadav et al. 1998); while black gram, mungbean and soybean are cultivated in early or mid-summer, followed by winter wheat (Hume, Shanmugasundaram, and Beversdorf 1985; Dhingra and Sekhon 1988; Rahman and Miah

TABLE 3. Areas and average national yields of common grain legumes grown in South Asia.

Species	Area ($\times$ 10^3 ha)	Average Yield (kg/ha)
Bangladesh[1]		
Lathyrus	226	815
Lentil	206	790
Chickpea	84	593
Black gram	64	765
India[2]		
Chickpea	6888	735
Black gram	3335	477
Mungbean	3302	431
Pigeonpea	3256	756
Nepal[3]		
Lentil	162	699
Lathyrus	26	543
Pigeonpea	26	729
Black gram	27	654
Soybean	21	731
Pakistan[4]		
Chickpea	1045	393
Lathyrus	138	536
Mungbean	168	411
Black gram	65	446

[1]Bangladesh Bureau of Statistics, 1999.
[2]Bhan & Mishra, 1998.
[3]Anonymous 1998
[4]Government of Pakistan, 1995.

1988). Recent breeding progress with short duration pigeonpea has made it possible to grow pigeonpea either as a summer substitute for rice or as a winter crop in rice-fallow cropping systems (Ali 1990).

It is interesting to note that lathyrus (*Lathyrus sativus*) is one of the leading pulse crops in terms of acreage in most of the countries. As both a grain and forage crop, lathyrus is well known for its relative disease and insect resistance as well as its ability to grow well under adverse soil and climatic conditions (low soil fertility, flood, drought) (Deshpande and Deshpande 1991). The popularity of lathyrus in Bangladesh, Nepal, and Pakistan speaks to the increasing, poor production by preferred grain legumes, the abundance of poor soils, and farmers' desire to grow some kind of pulse crop. It is ironic; however, that this easy to grow crop is also the source of a neurotoxin that can cause lathyrism, a permanent, disabling, neurological disorder. Some progress has been made towards reducing the neurotoxin, 3-oxalyl-L-2, 3-diaminopropanoic acid (β-ODAP), through plant breeding efforts (Mehra, Raju, and Himabindu 1995; Shaikh, Majid, and Lahiri 1995). Also a recently identified inverse relationship between soil zinc levels and β-ODAP content suggests a possible management practice to reduce the toxin (Lambein et al. 1994). Nevertheless lathyrism is still a concern for people and livestock who consume large quantities of the grain and foliage.

Average national yields of the leading grain legume species in South Asia are low (Table 3). Generally grain legumes do not have the potential for the same level of productivity as cereals, because of physiological energy costs which reduce yields. These costs include N fixation (10% reduction in yield), protein generation and photorespiration (30% reduction in production) (Phillips 1993). In addition to the inherent characteristics that keep legume yield potentials less than cereals, major gaps also exist in South Asia between optimum experimental yields and average national yields (Tables 3, 4). The biotic, abiotic and socioeconomic constraints responsible for these gaps will be discussed in the next section

Green Manure and Forage Legumes

Legumes utilized as green manures in rice-wheat cropping systems can be broadly characterized as pre-rice or post-rice. Species such as *Crotalaria juncea, Sesbania aculeata, S. rostrata, Cyamopsis tetragonoloba and Vigna unguiculata* have been used as 45-60 day catch

TABLE 4. Optimum experimental yields for recently released popular grain legumes grown in South Asia.

Species	Optimum Experimental Yields (kg ha^{-1})			
	Bangladesh	India	Nepal	Pakistan
Black gram	1500-1700[1]	1360-1760[2]	1000-1500[3]	NA
Chickpea	1400-3100[4]	1500-3200[5,6]	2000-3400[7]	1559-3300[8,9]
Lathyrus	1310-1740[10]	NA	1000-1700[11]	NA
Lentil	2300-3470[12,13]	1700[14]	1500-4100[7]	3843[15]
Mungbean	968-1108[16]	1000-1500[17]	1500[7]	1021-1500[18]
Pigeonpea	NA	2650-4570[19,20]	2000-2700[7]	NA
Soybean	NA	3500-4500[21]	1800-3600[7]	NA

NA = not available; [1] BARI 1998; [2] Singh and Srivastava 1988; [3] Regmi 1990; [4] Islam et al. 1991; [5] Bahl 1988; [6] Saxena and Johansen 1990; [7] ACD 1997; [8] Haq et al. 1999; [9] Ali 1999; [10] Rahman et al. 1991; [11] NARC 1998; [12] Sarkar et al. 1999; [13] Sarkar et al. 1992; [14] Solanki et al. 1993; [15] Tufail et al. 1995; [16] Shaikh et al. 1988; [17]Tickoo and Jain 1988; [18] Malik 1987; [19] Singh 1988; [20] Wallis et al. 1983; [21] Bhatnagar and Ali 1995.

crops in the pre-rice phase between wheat and rice (Yadav et al. 1998; Zahid, Mann, and Shah 1998). Biomass from these legumes is incorporated into the soil at the onset of the monsoon season, just prior to rice transplanting. Post-rice species include forage legumes *Stylosanthes, Trifolium alexandrinum, Trifolium repens, Vicia bengalensis, Vicia villosa, Clitoria ternatea, Desmanthus virgatus, Cyamopsis tetragonoloba,* and *Macroptilium atropurpureum* which are substituted for wheat or other winter crops (Meelu and Rekhi 1981; De, Rao, and Ali 1983; Bhattarai 1996; Ladha et al. 1996; Schulz 1997). *Lathyrus sativus* is cultivated in Pakistan during the winter months primarily as a fodder crop (Zahid, Mann, and Shah 1998).

CONSTRAINTS TO PRODUCTIVITY

Diversifying the rice-wheat cropping system in South Asia with grain, forage or green manure legumes is a reality, but at too small a scale to significantly impact human and livestock diets or to improve the soil resource base for rice and wheat. Poor and uncertain productivity makes cultivation of legume crops an option that resource-poor, risk-adverse farmers are unlikely to adopt for either health or long range sustainability reasons. Lack of knowledge about the health benefits of legumes in cereal-based diets may also contribute to the lack of enthusiasm for growing legumes.

Proceedings from conferences convened in the 1980's and 90's thoroughly elucidate the constraints to high yields with grain legumes (Summerfield and Roberts 1985; Wallis and Byth 1987; Baldev, Ramanujam, and Jain 1988; Muehlbauer and Kaiser 1994). A recent workshop hosted by the International Center for Research in the Semi-Arid Tropics (ICRISAT) specifically examined the constraints and opportunities for grain legumes in rice-wheat cropping systems of the Indo-Gangetic Plain (Johansen et al. 2000). Table 5 summarizes the disease; insect and weed pests identified as major biotic problems for the leading grain legumes in the region.

The most pervasive of the diseases appear to be: yellow mosaic virus in black gram and mungbean; botrytis gray mold (*Botrytis cenerea*) in lentil and chickpea; ascochyta blight (*Ascochyta rabiei, Ascochyta lentis*) in chickpea, lentil and mungbean; cercospora leaf spot (*Cercospora cruenta*) in black gram, mungbean and soybean as well as wilt (*Fusarium oxysporum*) of chickpea, lentil and pigeonpea. Estimates of yield reductions from such diseases range from 42-90%, 10-70%, 50%, and 12-50% in Bangladesh, India, Nepal, and Pakistan, respectively (Rahman et al. 2000; Ali et al. 2000; Pandey et al. 2000; Haqqani and Zahid 2000).

Likewise major insect pests are considered to be hairy caterpillar (*Diacrisia obliqua*) in black gram, mungbean and soybean as well as various pod borers (*Helicoverpa armigera, Maruca testulalis, Euchrysops cnejus*) of black gram, chickpea, lentil and pigeonpea. Yield reductions from these pests also are high, varying from 15% to 90% across South Asia (Rahman et al. 2000; Ali et al. 2000; Pandey et al. 2000; Haqqani and Zahid 2000).

Weeds are serious constraints to grain legume production, yet only a few species have been identified in India and Nepal namely, lambsquarters (*Chenopodium album*), purple nutsedge (*Cyperus rotendus*) and bermudagrass (*Cynodon dactylon*). Similarly parasitic nematodes such as root knot (*Meliodogyne incognita, Meliodogyne javanica*), cyst (*Heterodera cajani*), lesion (*Pratyenchus* spp.) and stunt (*Tylenchorynchus* spp.) have been described in India, but not in the other countries. Yield reductions from weeds and nematodes are estimated at 10-80% and 12-19%, respectively (Rahman et al. 2000; Ali et al. 2000; Pandey et al. 2000; Haqqani and Zahid 2000).

Abiotic constraints to good legume production reflect sub-optimal management and the secondary status of legumes relative to cereals. Unlike wheat that usually receives 2-3 irrigations, winter legumes

TABLE 5. Summary of biotic constraints to increased grain legume production in South Asia.

DISEASES	INSECTS	NEMATODES	WEEDS
Black gram			
Yellow mosaic virus (B, I, N, P) Cercospora leaf spot (B, I, N) Powdery mildew (B, I) Pod/leaf blight (N) Leaf crinkle virus (P)	Hairy caterpillar (B, N, P) Aphids (N) Whitefly (B) Helicoverpa (B) Blue butterfly pod borer (B) Monolepta beetle (B) Cut worm (B)	*Meloidogyne incognita* (I) *M. javanica* (I)	Significant but unspecified (B, I, N, P)
Chickpea			
Botrytis gray mold (B, I, N) Ascochyta blight (I, P) Wilt (B, P) Root/collar rot (B, I)	Helicoverpa (B, I, N, P) Cutworm (B, I) Semilooper (I, P)	*Meloidogyne incognita* (I) *M. javanica* (I) *Pratyenchus* spp. (I) *Tylenchorynchus* spp. (I)	*Chenopodium album* (I) *Cyeprus rotendus* (I) *Cynodon dactylon* (I) Significant but unspecified (B, P)
Lathyrus			
Powdery mildew (B, N)	Aphids (B, N)	--	Significant but unspecified (P)
Lentil			
Stemphylium blight (B, I) Rust (B, I, N) Wilt (I, N) Botrytis gray mold (I, N) Ascochyta blight (I, P) Powdery mildew (I)	Helicoverpa (B) Cutworm (B) Aphid (I, N) Etiella pod borer (I)	*Meliodogyne incognita* (I) *M. javanica* (I)	*Chenopodium album* (I, N) *Cyperus rotendus* (I, N) *Vicia sativa* (N) Significant but unspecified (B)
Mungbean			
Yellow mosaic virus (I, N, P) Cercospora leaf spot (I, N, P) Ascochyta blight (I, P) Powdery mildew (I) Pod/leaf blight (N, P)	Hairy caterpillar (P) Aphids (N) Whitefly (I) Leaf hopper (I) Thrips (I)	*Meloidogyne incognita* (I) *M. javanica* (I)	Significant but unspecified (B, I, N, P)
Pigeonpea			
Sterility mosaic disease (I, N) Phytophthora blight (I) Alternaria leaf spot (I) Wilt (I, N) Root/collar rot (N) Stem canker (N)	Helicoverpa (N, I) Spotted pod borer (I) Blue butterfly pod borer (I) Pod fly (I) Blister beetle (I)	*Heterodera cajani* (I)	Significant in early stages but unspecified (I, N)
Soybean			
Cercospora leaf spot (N) Pod/stem blight (N) Rust (N)	Hairy caterpillar (N)	--	Significant but unspecified (N)

Occurrence: B = Bangladesh; I = India; N = Nepal; P = Pakistan; Adapted from Johansen et al. 2000.

often are expected to grow on residual soil moisture and rainfall. Inadequate winter precipitation or late planting in soils with insufficient moisture reserves often leads to drought stress at critical plant growth stages (Ali et al. 2000). On the other hand, precipitation early in the cropping season may leave soils waterlogged, thereby preventing good land preparation and stand establishment. Excess moisture later in the season results in too much vegetative growth and problems with diseases and insects (Pandey et al. 2000). Temperatures above 30°C often accompany droughty conditions, reducing vegetative growth and causing flower abortion or slower pod filling (Rahman et al. 2000). Low winter temperatures in some parts of the Indo-Gangetic Plain retard legume vegetative growth; increase competition with weeds and cause flower drop (Haqqani and Zahid 2000).

As discussed in other chapters of this volume, soil constraints are a major factor regarding the sustainability of rice and wheat production in South Asia. The once highly fertile soils of the Indo-Gangetic Plain now have low fertility; and soil organic matter levels less than 1% are common. Given this situation on the "best" lands, it is not surprising that legume cultivation in the region suffers from soil nutrient problems. Farmers do not generally apply fertilizers to legumes, expecting the crop to grow on low, native soil fertility reserves. Yet significant legume yield responses to N, P and K as well as micronutrients B, Zn and Mo have been reported throughout the region (Mondal et al., 1991; Karim, Miah, and Hassain 1992; Kushwaha 1993; Sakal 1993; Arif et al. 1996; Nagaraju and Yadahalli 1996; Pasricha and Bahl 1996; Sakal et al. 1998; Singh and Singh 1998; Gupta and Satinder 1999; Rashid, Din, and Bashir 1999; Singh 1999; Srivastava et al. 1999). In addition to low soil pH, saline-sodic soils, flooding and nutrient deficiencies adversely affect nitrogen fixation processes, thereby preventing the short- and long-term agronomic benefits of legumes from reaching their potential (Sharma, Minhas, and Masand 1988; Peoples and Craswell 1992). Finally many of the soil physical properties that constrain good wheat productivity after rice (restricted rooting zones, poor infiltration, low water holding capacity) also apply to legumes, leading to poor plant stands, lodging and vulnerability to drought stress (So and Woodhead 1987).

Socioeconomic limits to increased legume production are quite similar in Bangladesh, India, Nepal, and Pakistan. Some of the more important issues include:

Cost-benefit ratios–Despite lower cultivation costs for grain legumes relative to rice-wheat, and higher market prices for legumes, farmers consider grain legumes low economic return crops. The cost-benefit ratio of pulse crops is less attractive as compared with that of rice and wheat, primarily because of poor legume yields. So while the output price of pigeonpea is double that of rice, the yield level of rice is four times higher than pigeonpea (Ali et al. 2000).

Low output price at harvest–Most farmers do not have storage facilities, so growers must sell grain legumes at the market soon after harvest. Prices tend to be very low at harvest time, then rise steeply soon afterwards. As a result, the margin between prices paid by consumers and those received by the farmer is very high (Joshi 1998). Non-remunerative prices at the farmers' level are a major disincentive and are considered responsible for lowering pulse production overall. Government policies for proper marketing systems and fair farm-gate prices are an urgent need.

Poor economic condition of the farmers–Most farmers are economically below the poverty level and own very small land holdings. The input-mobilization power of the poor farmer is very limited, primarily because of high input costs like quality seeds, fertilizers, pesticides, etc. In addition timely availability and quality of these inputs is not always assured.

Poor linkages between farmers and extension service–At present, there are few, if any, extension service programs for increasing legume cultivation. Suitable technology packages along with strong technology transfer efforts are necessary to increase legume production.

Lack of appropriate policy support–It is well known that current government policies in the region emphasize cereal production over legumes. Institutional credit, subsidies for inputs and crop failure, provision for support service, etc., are very much lacking for pulses.

ADDRESSING POOR PRODUCTIVITY

Despite the heavy burden of constraints, there is hope for increased legume production in South Asia. Crop improvement efforts are responsible for this optimism. Over the last 25 years, plant breeders in the international centers ICRISAT, ICARDA and AVRDC as well as the Bangladeshi, Indian, Nepali, and Pakistani national programs have focussed on increasing grain legume yield potentials and improving tolerance or resistance to major insects and diseases. Table 4 indicates the gains made in yield potentials, while Table 6 details some of the recent grain legume varietal releases aimed at pests and diseases. These cultivars provide environmentally friendly and accessible options for resource-poor farmers instead of chemical pest and disease products that can cause pollution and are often beyond the financial resources of farmers.

Such genetic advances need to be strongly coupled to extension and the legume seed industry, so that farmers are adequately informed about the capabilities of improved varieties, and that seed supplies are readily available in response to increased adoption. Such factors led to the successful and expanded adoption of improved pigeonpea and chickpea varieties in some sectors of India (Bantilan and Parthasarathy 1998; Joshi et al. 1998). However, in the Terai region of Nepal, adoption of improved chickpea varieties Sita, Dhanush, and Kalika (Table 6) were only 10-37%, 32%, and 14% of the area, respectively (Gurung 1998). Much of the slow adoption was attributed to the absence of a legume seed industry and dissemination infrastructure. Approximately 9 to 103 t of improved chickpea seed were required by farmers in this area of Nepal; however, the capacity of the government institution responsible for distributing legume seed was only 2-3 t. A similar situation exists in Bangladesh (Hussain 1991). These conditions are not promising for promoting legume production in Nepal and Bangladesh.

Good agricultural practices must be integrated with improved germplasm in order for legume productivity to reach its potential. A few management approaches that have promise for synergistic effects with improved germplasm follow.

Land Preparation

Farmers generally invest little in land preparation for legume planting, leaving seeds exposed to bird and insect scavenging as well as to extreme heat and dry soil conditions. As a result, low germination, poor stand establishment and high weed pressure are common (McWilliam and Dillon 1987). The increasing popularity of power tiller units in Bangladesh, parts of India and Nepal provide new opportunities for improved land preparation and seeding of legume crops. Preliminary efforts by national agricultural scientists (NARS) in Nepal and Bangladesh have shown that seeding with a power tiller-cum-seeder ensures uniform seed placement at the appropriate depth (SM-CRSP

TABLE 6. Improved grain legume varieties released through South Asian national programs.

BANGLADESH	INDIA	NEPAL	PAKISTAN
		Chickpea	
Hyprosola, Sabur, Nabin, Barichola2 (W, RR), Barichola3 (W), Barichola4 (W), Barichola5 (W,BGM) Barichola6-8(W), Binasola2(W)	GL83119, 84038, 84096, 84107(AB, W), H86-84, 86-18 (W, RR), BG276, ICC1069 (BGM), PDE2, PDG84-10 (PB), K1122, BG302, IG218, GCP 11 (RKN)	Dhanush (BGM), Trishul, Sita, [ICCC4] Radha [JG74] (W) Kalika [ICCL82108] (W) Koselee [ICCC32], ICCV6 (W, PB)	CM 98 (AB, W), AUG480 (AB, W) CM72, CM88 (AB), DG89 (PB, RR) DG92, Noor91, Paidar91. Punjab91 (AB), NIFA-88 (AB)
		Lentil	
Utfala, Barimasur-1 (SB, R), Barimasur-2 (SB, R), Barimasur-3 (SB, R), Barimasur-4 (SB, R)	PL406, PL639®, LL30, 56, 78, 112, 116 ®, UPL175, PL81-17, L1304, DPL16 (W, RR), PL 639, LL301, LG 178 (AB)	Sishir, Sindur, Simrik, Simal, Sikhar ®, Khajura 1, Khajura 2	Mansoor85 (AB, R), Masoor93 (AB, BGM, R, SR) Manserha89 (AB, R)
Mungbean	Mungbean	Pigeonpea	Mungbean
Mubarik, Kanti, Barimung2-5 (CLS, YMV), Binamoog1-5 (CLS, YMV)	Pusa101,105 (YMV, PM) ML131, 267, 337 (YMV) Sujata, Dhauli Mut. (CLS, PM) Pant M1, 2, 3 (YMV) Gujarat2 (CLS)	ICPL366 (W, SMD) Rampur Rahar (W, SMD) ICPL87133 (W, SMD) ICPL84072 (W, SMD), ICP11384	NIAB Mung 92 (YMV) NIAB-M51, -M54 (YMV, CLS) NM 19-19, 121-25, 13-1, 20-21 (YMV)
Lathyrus	Pigeonpea	Soybean	Blackgram
Barikhesari-1	ICPL85010 (SMD, PB) ICPM 93003 (W, SMD) ICPL87162 (SMD) ICPL87119 (W, SMD) BSMR736 (W, SMD) BSMR175 (SMD)	Seti, Lumle-1	Mash-1, 2, 3 (YMV)

R = Rust, SMD = Sterility mosaic disease, AB = Ascochyta blight, YMV = Yellow mosaic virus, W = Wilt, CLS = Cercospora leaf spot, RR = Root rot, SR = Stem rot, PM = Powdery mildew, SB = Stemphylium blight; PB = Pod borers, BGM = Botrytis gray mold, RKN = Root knot nematode.

2000). Additional research is needed to quantify the benefits on legume germination, stand establishment and yields. Although economic assessments of this technology for legumes have not yet been performed, similar studies for wheat have shown the power seeder option to be more profitable than traditional animal traction (Hasan, Miah, and Mandal 1991; Hossain and O'Callaghan 1996).

Seedling Vigor

In addition to planting stresses, high temperatures and humidity, oil content, insects, pathogens and mechanical damage contribute to lower viability and vigor of legume seeds (Heslehurst, Imrie, and Butler 1987). Recent work by Harris et al. (1999) and Musa et al. (1999) shows that a simple water priming procedure of farmer chickpea seeds was able to overcome some of the loss in viability and vigor. Seeds were soaked in water overnight, surface dried and immediately planted. Earlier chickpea seedling emergence, better plant stands and seedling vigor along with increases in grain yield of 25-67% and 47% were observed on multiple farmer fields in India and Bangladesh, respectively. Also less soil-borne diseases and pod borer damage were found on the primed compared to unprimed seedlings (Musa et al. 1999). Even more encouraging was the finding that some farmers used fertilizer on their primed crops because they perceived less risk (Harris et al. 1999).

Soil Fertility

Research and on-farm trials in the region have shown that legume crop establishment and yields are responsive to phosphorus fertilizer applications (Kushwaha 1993; Singh et al. 1994; Nagaraju and Yadahalli 1996). Phosphorus also shows benefits in terms of increased nodulation (Jan et al. 1994; Singh, Singh, and Singh 1995), and disease resistance (Sharma, Verma, and Chauhan 1996; Singh and Chand 1996). Likewise deficiencies of lime in acid soils and micronutrients such as Zn, B, and Mo are a major cause of low grain legume yields in some parts of South Asia (Craswell, Loneragan, and Keeratai-Kasikorn, 1987; Srivastava et al. 1999).

While legume crops are often recommended as a low cost, sustainable means for improving soil fertility, few benefits are likely, if the legume crops themselves do not grow well. However, affordability, accessibility and fertilizer quality often preclude critical fertilizer input use by farmers. Strong extension networks, coupled with farmer participatory schemes, and concerted government intervention appear to be the best approaches for addressing these constraints (Gowda, Faris, and Maniruzzaman 1994). Government interventions could include fertilizer subsidies, low rate credit facilities, price supports and fair market price policies as well as fertilizer quality certification.

Weeds

As discussed earlier, unchecked weed growth in legume crops is a common situation and a cause of significant yield reductions. Row planting and mulching are low cost and environmentally benign methods for overcoming weeds, whereas numerous herbicides offer more capital intensive options. Row planting allows efficient mechanical cultivation by hoe or other means, relative to broadcast seeding (Sharma and Thakur 1993). Likewise mulching with wheat or rice straw shades out weeds, while conserving valuable soil moisture (Kolar, Grewal, and Singh 1979; Hamada et al. 1991).

Forage and Green Manure Legume Value

Many of the technical problems detailed for grain legumes also apply to forage and green manure legumes. In addition green manures or forage legumes are often perceived as crops with no economic return, because of competition with cereal production. This is a logical approach given the importance of supplying enough food for rapidly expanding populations in the region. However, it should be recognized that a thorough analysis of the positive and negative trade-offs of increasing forage or green manure legume acreage has not been undertaken. Some questions that might be asked are: Would increased animal productivity change the quantity/quality of manures applied to fields for maintenance of soil fertility? What effects would increased rice or wheat production after forage legumes or green manure crops have on residue supply? Would more residues be available for soil incorporation and soil fertility improvement? How might increased income from better animal productivity or fewer purchased inputs (concentrates, N fertilizer) influence the decisions that farmers make?

Understanding the impacts of forage or green manure legumes from a systems perspective may be helpful in encouraging adoption or lead to government policies that would promote legume cultivation.

CONCLUSIONS

High yielding rice and wheat crops are critical for food security in South Asia. We believe that increased cultivation of grain, forage and green manure legumes are equally critical for regenerating degraded/ exhausted soils, and for providing essential protein, minerals and vitamins to needy humans and livestock in the region. There is ample scope for increased legume production within the rice-wheat cropping system without jeopardizing food security. However, in order to realize sustainability and human/livestock health impacts, existing production of legumes must increase significantly. Promising genetic and agronomic options have been presented for increasing yields, with the hope of encouraging farmers to grow legumes more frequently within the rice-wheat system.

Much additional work still remains to ensure increased legume productivity including the identification of genotypes resistant to more insect pests, tolerant to drought or waterlogging stresses and competitive with weeds. More importantly strong government commitments, investments and institutional support are essential, if legume production is to increase and if South Asian farmers are to be persuaded to invest in legumes.

REFERENCES

ACD. (1997). *Agriculture Diary*. Ministry of Agriculture, Agriculture Communication Division. Lalitpur, Nepal.

Ahlawat, I.P.S., M. Ali, R.L. Yadav, J.V.D.K. Kumar Rao, T.J. Rego, and R.P. Singh. (1998). Biological nitrogen fixation and residual effects of summer and rainy season grain legumes in rice and wheat cropping systems of the Indo-Gangetic Plain. In *Residual Effects of Legumes in Rice and Wheat Cropping Systems of the Indo-Gangetic Plain*, eds. J.V.D.K. Kumar Rao, C. Johansen, and T.J. Rego. New Delhi and Pantancheru, India: Oxford & IBH Publishing Co. Pvt. Ltd. and ICRISAT, pp. 31-54.

Ahuja, A.K., B.K. Gupta, M.S. Sohoo and B.L. Bhardwaj. (1991). Comparative nutritive value of maize fodder grown with and without cowpea. *Indian Journal of Dairy Science* 44(6):389-391.

Akbar, M.A. and P.C. Gupta. (1990). Nutritive value of faba bean (*Vicia faba* l.) seeds, fodder and silage. *FABIS Newsletter* 26: 38-41.

Ali, A. (1999). Bittal 98 (A 16): An improved form of the most predominant desi chickpea variety C 44. *International Chickpea and Pigeonpea Newsletter* 6: 8-9.

Ali, M. (1990). Pigeonpea: Cropping systems. In *Pigeonpea*, eds. Y.L. Nene, S.D. Hall, and V.K. Sheila. Wallingford, England: CAB International for ICRISAT, pp. 279-301.

Ali, M., P.K. Joshi, S. Pande, M. Asokan, S.M. Virmani, R. Kumar, and B.K. Kandpal. (2000). Legumes in the Indo-Gangetic Plain of India. In *Legumes in Rice and Wheat Cropping Systems of the Indo-Gangetic Plain-Constraints and Opportunities*. eds. C. Johansen, J.M. Duxbury, S.M. Virmani, C.L.L. Gowda, S. Pande, and P.K. Joshi. Pantancheru, India and Ithaca, NY: ICRISAT and Cornell University. In press.

Anonymous. (1998). *Statistical Information on Nepalese Agriculture*. Agricultural Statistics Division Ministry of Agriculture, Kathmandu, Nepal.

Arif, I., Samiuliah, M.M.R.K. Afridi, and U. Shahid. (1996). Potassium nutrition under different irrigation levels in selected crops. *Journal of Potassium Research* 12(2): 186-193.

Aykroyd, W.R., J. Doughty, and A. Walker. (1982). Legumes in human nutrition. *FAO Food and Nutrition Paper* No. 20, Rome, pp. 1-152.

Babu, S. (1976). *Studies on Folic Acid*. PhD Thesis of Osmania University.

Bahl, P.N. (1988). Chickpea. In *Pulse Crops*, eds. B. Baldev, S. Ramanujam, and H.K.Jain, New Delhi, India: Oxford and IBH Publishing Co. Pvt. Ltd., pp. 95-131.

Baldev, B. (1988). Origin, distribution, taxonomy and morphology. In *Pulse Crops*, eds. Baldev, B., S. Ramanujam, and H.K. Jain, New Delhi, India: Oxford and IBH Publishing Co. Pvt. Ltd., pp. 3-51 Oxford and IBH Publishing Co. Pvt. Ltd. 626 p.

Bamji, M.S. and A.V. Lakshmi. (1998). Less recognized micronutrient deficiencies in India. *Nutrition Foundation of India Bulletin* 19(2): 5-8.

Bangladesh Agricultural Research Institute (1998). *Lentil, Mungbean and Black Gram Pilot Project*. Publications 5, 15 and 16.

Bangladesh Bureau of Statistics. (1999). *Monthly Statistical Bulletin of Bangladesh*, Jan.-July 1999.

Bantilan, M.C.S. and D. Parthasarathy. (1998). Adoption assessment of short-duration pigeonpea ICPL 87. In *Assessing Joint Research Impacts: Proceedings of an International Workshop on Joint Impact Assessment of NARS/ICRISAT Technologies for the Semi-Arid Tropics*, 2-4 Dec. 1996, eds. M.C.S. Bantilan and P.K. Joshi, ICRISAT, Patancheru, India, pp. 136-152.

Becker, M., J.K. Ladha, and M. Ali. (1995). Green manure technology: Potential, usage, and limitations–A case study for lowland rice. *Plant and Soil* 174: 181-194.

Benbi, D.K. and S.P.S Brar. (1992). Dependency of DTPA-extractable Zn, Fe, Mn, and Cu availability on organic carbon presence in arid and semiarid soils of Punjab. *Arid Soil Research and Rehabilitation* 6: 207-216.

Beri, V., O.P. Meelu, and C.S. Khind. (1989b). Studies on *Sesbania aculeata* Pers. as green manure for N accumulation and substitution of fertilizer N in wetland rice. *Tropical Agriculture* 66: 209-212.

Bhan, V.M. and J.S. Mishra. (1998). Efficient use and management of agro-chemicals for increasing pulse production. *Agricultural Situation in India*, 187-196.

Bhardwaj, S.P., S.N. Prasad, and G. Singh. (1981). Economizing nitrogen by green manures in rice-wheat rotation. *Indian Journal of Agricultural Science* 51: 86-90.

Bhatnagar, P.S. and N. Ali. (1993). Country Report 4-India. In *Soybean in Asia*, eds. N. Chomchalow and P. Laosuwan, RAPA Publication:1993/6, Lebanon, NH: Science Publishers, Inc., pp. 34-49.

Bhattarai, S. (1996). Biological nitrogen fixation and residual effects of winter legumes-Nepal experiences. Paper presented to the *Workshop on Residual Effects of Legumes in Rice and Wheat Cropping Systems of the Indo-Gangetic Plain*. 26-28 August 1996. ICRISAT Asia Center, Pantancheru, India.

Boparai, B.S., Yadvinder-Singh, and B.D. Sharma. (1992). Effect of green manuring with *Sesbania aculeata* on physical properties of soil and on growth of wheat in rice-wheat and maize-wheat cropping systems in a semiarid region of India. *Arid Soil Research and Rehabilitation* 6(2): 135-143.

Bressani, R., and L.G. Elias. (1974). Legume foods. In *New Protein Foods*, ed. A.M. Altschul, New York, NY, Academic Press, pp. 230-297.

Bressani, R., and L.G. Elias. (1978). Nutritional value of legume crops for humans and animals. In *Advances in Legume Science*, eds. R.J. Summerfield and A.H. Bunting, Kew, England: Royal Botanic Gardens, pp. 135-155.

BRRI (Bangladesh Rice Research Institute). (1985). *Annual Report for 1984*. Bangladesh Rice Research Institute, Joydebpur, Gazipur.

Buresh, R.J. and S.K. DeDatta. (1991). Nitrogen dynamics and management in rice-legume cropping systems. *Advances in Agronomy* 45: 1-59.

Carangal, V.R., E.T. Rebancos, E.C. Armada, and P.L. Tengco. (1994). Integration of forage and green manure production systems. pp. 51-67. In *Green Manure Production Systems for Asian Ricelands*. eds. J.K. Ladha and D.P. Garrity, Int. Rice Res. Inst., Los Banos, Philippines.

Chang, R., S. Schwimmer, and H.K. Burr. (1977). Phytate: Removal from whole dry beans by enzymatic hydrolysis and diffusion. *Journal of Food Science* 42: 1098.

Chatterjee, B.N., K.I. Singh, A. Pal, and S. Maiti. (1979). Organic manures as substitutes for chemical fertilizers for high-yielding rice varieties. *Indian Journal of Agricultural Science* 49: 188-192.

Chatterjee, S.R. and Y.P. Abrol. (1975). Amino acid composition of new varieties of cereals and pulses and nutritional potential of cereal-pulse combination. *Journal of Food Science and Technology*. 12: 221-227.

Chaudhary, S.L. (1990). Response of different sources of nitrogen fixing green manures on yield of paddy-wheat at Tarhara during 1988-89. *Nitrogen Fixing Tree Research Reports* 8: 48-50.

Chwang L.C., A.G. Soemantri, and E. Pollitt. (1988). Iron supplementation and physical growth of rural Indonesian children. *American Journal of Clinical Nutrition* 47 (3): 496-501.

Craswell, E.T., J.F. Loneragan, and P. Keerati-Kasikorn. (1987). Mineral constraints to food legume crop production in Asia. In *Food Legume Improvement for Asian Farming Systems*: *Proceedings of an International Workshop*, eds. E.S. Wallis and

D.E. Byth, 1-5 September 1986, Khon Kaen, Thailand. ACIAR Proceedings No. 18. Canberra, Australia: ACIAR, pp. 99-111.

Dargan, K.S., R.K. Chillar, and K.K.R. Bhardwaj. (1975). Green manuring for more paddy. *Indian Farming* 25: 13-15.

De, R., Y.Y. Rao, and W. Ali. (1983). Grain and fodder legumes as preceding crops affecting the yield and N economy of rice. *Journal of Agricultural Science* 101: 463-466.

DeDatta, S.K. and S.S. Hundal. (1984). Effects of organic matter management on land preparation and structural regeneration in rice-based cropping systems. In *Organic Matter and Rice*. Int. Rice Res. Inst., Los Banos, Philippines, pp. 399-416.

Deshpande, S.S., B. Singh, and U. Singh. (1991). Cereals. In *Foods of Plant Origin*. eds. D.K. Salunkhe and S.S. Deshpande, New York, NY: Van Nostrand Reinhold, pp. 6-136.

Deshpande, S.S., S.K. Sathe, and D.K. Salunkhe. (1984). Dry beans of *Phaseolus*: A review. Part 3. *CRC Critical Reviews in Food Science and Nutrition* 21(2): 137-195.

Deshpande, U.S., and S.S. Deshpande. (1991). Legumes. In *Foods of Plant Origin*. eds. D.K. Salunkhe and S.S. Deshpande, New York, NY: Van Nostrand Reinhold, pp. 137-300.

Devendra, C. (1997). Crop residues for feeding animals in Asia: Technology development and adoption in crop/livestock systems. In *Crop Residues in Sustainable Mixed Crop/Livestock Farming Systems*, ed. C. Renard, CAB International, NY, NY, pp. 241-267.

Dhingra, K.K., and H.S. Sekhon. (1988). Agronomic management for high productivity of mungbean in different seasons, Punjab, India. In *Mungbean: Proceedings of the Second International Symposium*. 16-20 November 1987, Bangkok, Thailand. Asian Vegetable Research and Development Center (AVRDC), Shanhua, Tainan: AVRDC, pp. 378-384.

Dilawari, J.B., P.S. Kamath, R.P. Batta, S. Mukewar and S. Raghavan (1981). Reduction of postprandial plasma glucose by Bengal gram dal (*Cicer arietinum*) and Rajmah (*Phaseolus vulgeris*). *American Journal of Clinical Nutrition* 34: 2450-2453.

FAOSTAT. (1999). Food and Agricultural Organization (FAO) Statistical Databases, Rome, Italy: FAO. http://apps.fao.org/cgi-bin/nph-db.pl

Food and Agricultural Organization (FAO) (1993). *Rice in Human Nutrition*, Food and Nutrition Series No. 26. FAO, Rome, pp. 1-162.

Food and Agricultural Organization and World Health Organization. (1973). *Report on Energy and Protein Requirements*, Joint FAO/WHO Ad Hoc Expert Committee on Energy and Protein Requirements. World Health Organization Technical Report Series No. 52. and FAO Nutrition Meetings Report Series No. 52, 118 p.

Food and Agricultural Organization. (1998a). *Nutrition Country Profile: Pakistan*, Rome, Italy: FAO.

Food and Agricultural Organization. (1998b). *Nutrition Country Profile: India*, Rome, Italy: FAO.

Food and Agricultural Organization. (1999). *Nutrition Country Profile: Bangladesh*, Rome, Italy: FAO.

Garrity, D.P. and J.C. Flinn. (1988). Farm-level management systems for green manure crop in Asian rice environments. In *Green Manure in Rice Farming*. Los Banos, Philippines: International Rice Research Institute, pp. 111-129.

Gill, M.S., T. Singh, and D.S. Rana. (1994). Integrated nutrient management in rice (*Oryza sativa*)-wheat (*Triticum aestivum*) cropping sequence in semi-arid tropic. *Indian Journal of Agronomy* 39(4): 606-608.

Golden, M.H.N., and B.E. Golden. (1981). Effect of zinc supplementation on the dietary intake, rate of weight gain, and energy cost of tissue deposition in children recovering from severe malnutrition. *American Journal of Clinical Nutrition* 34: 900-908.

Gopalan, C. (1998). Micronutrient malnutrition in SAARC–The need for a food-based approach. *Nutrition Foundation of India Bulletin*, 19(3):1-4.

Goswami, N.N., R. Prasad, M.C. Sarkar, and S. Singh. (1988). Studies on the effect of green manuring in nitrogen economy in a rice-wheat rotation using a ^{15}N technique. *Journal of Agricultural Science* 111: 413-417.

Government of Pakistan (1995). *Pakistan Statistical Yearbook*, Central Statistical Office, Statistics Division.

Gowda, C.L.L., D.G. Faris, and A.F.M. Maniruzzaman. (1994). Infrastructural support to promote farmer adoption of improved technologies. In *Expanding the Production and Use of Cool Season Food Legumes–Proceedings of the Second International Food Legume Research Conference on Pea, Lentil, Faba Bean, Chickpea, and Grasspea*, eds. F.J. Muehlbauer and W.J. Kaiser, 12-16 April 1992, Cairo, Egypt. Dordrecht, The Netherlands: Kluwer Academic Publishers.

Gupta, S.P. and D. Satinder. (1999). Comparative response of some rabi crops to zinc application in Ustipsamment soil of Haryana. *Annals of Agriculture and Biological Research* 4(2): 157-160.

Gurung, R.K. (1998). Improved chickpea varieties in Nepal. In *Assessing Joint Research Impacts: Proceedings of an International Workshop on Joint Impact Assessment of NARS/ICRISAT Technologies for the Semi-Arid Tropics*, eds. M.C.S. Bantilan and P.K. Joshi, 2-4 December 1996, Patancheru, India: ICRISAT. pp. 114-125.

Hamada, Y., T. Nonoyama, I. Shaku, and H. Kojima. (1991). Studies on no-tillage sowing cultivation of soyabeans. 3. Weed control method. *Report of the Tokai Branch of the Crop Science Society of Japan*. 111: 23-24.

Haq, M.A-ul., M. Sadiq, and M-ul. Hassan. (1999). CM 98 (CM 31-1/85): A very high-yielding, disease-resistant mutant variety of chickpea. *International Chickpea and Pigeonpea Newsletter* 6: 9-10.

Haqqani, A.M., and M.A. Zahid. (2000). Legumes in Pakistan. In *Legumes in Rice and Wheat Cropping Systems of the Indo-Gangetic Plain–Constraints and Opportunities*. eds. C. Johansen, J.M. Duxbury, S.M. Virmani, C.L.L. Gowda, S. Pande, and P.K. Joshi. Pantancheru, India and Ithaca, NY: ICRISAT and Cornell University. In press.

Harris, D., A. Joshi, P.A. Khan, P. Gothkar, and P.S. Sodhi. (1999). On-farm seed priming in semi-arid agriculture: Development and evaluation in maize, rice and chickpea in India using participatory methods. *Experimental Agriculture* 35: 15-29.

Hasan, M.K., M.T.H. Miah, and M.A.S. Mandal. (1991). Returns from power tiller utilization in Bangladesh: An analysis of farm level data. *Bangladesh Journal of Agricultural Economics*. 14: 87-97.

Heslehurst, M.R., B.C. Imrie, and J.E. Butler. (1987). Limits to productivity and adaptation due to preharvest and postharvest factors. In *Food Legume Improvement for Asian Farming Systems: Proceedings of an International Workshop*, eds. E.S. Wallis and D.E. Byth, 1-5 September 1986, Khon Kaen, Thailand. ACIAR Proceedings No. 18. Canberra, Australia: ACIAR, pp. 169-192.

Hossain, A.H.M.S. and J.R.A. O'Callaghan. (1996). A case study on the mechanization of wheat cultivation in Bangladesh. *Journal of Agricultural Engineering Research*. 65: 175-181.

Hume, D.J., S. Shanmugasundaram, and W.D. Beversdorf. (1985). Soyabean (*Glycine max* (L.) Merrill). In *Grain Legume Crops*. eds. R.J. Summerfield and E.H. Roberts, London, England: William Collins Sons & Co. Ltd., pp. 391-432.

Hussain, M.M. (1991). Pulses seed production and distribution program of the Bangladesh Agricultural Development Corporation. In *Advances in Pulses Research in Bangladesh-Proceedings of the Second National Workshop on Pulses*, 6-8 Jun. 1989, Joydebpur, Bangladesh. Bangladesh Agricultural Research Institute, International Development Research Centre and ICRISAT. Pantancheru, India: ICRISAT. pp. 157-161.

Islam, M.O., M.M. Rahman, A. Rahman, A. Sarkar, and W. Zaman. (1991). Status and future breeding strategies for chickpea. In *Advances in Pulses Research in Bangladesh-Proceedings of the Second National Workshop on Pulses*, 6-8 Jun. 1989, Joydebpur, Bangladesh. Bangladesh Agricultural Research Institute, International Development Research Centre and ICRISAT. Pantancheru, India: ICRISAT, pp. 11-17.

Jan, H., M. Subhan, M. Yaqoobi, R. Jan, and S.B. Khan. (1994). Effect of fertilization, *Rhizobium* inoculation and carbofuran on nodulation and yield of lentil. *Sarhad Journal of Agriculture* 10: 425-430.

Jansen, G.R. (1974). The amino acid fortification of cereals. In *New Protein Foods*, ed. A.M. Altschul, New York, NY, Academic Press, pp. 40-120.

Jaya, T.V., Venkataraman, L.V. and Krishnamurthy, K.S. 1979. Influence of germinated legumes on the levels of tissue cholesterol and liver enzymes of hyper-cholesterolemic rats: Influence of ungerminated and germinated chickpea and green gram (whole). *Nutritional Reports International* 20: 371-381.

Johansen, C., J.M. Duxbury, S.M. Virmani, C.L.L. Gowda, S. Pande, and P.K. Joshi. eds. (2000). *Legumes in Rice and Wheat Cropping Systems of the Indo-Gangetic Plain-Constraints and Opportunities*. Pantancheru, India and Ithaca, NY: ICRISAT and Cornell University. In press.

Joshi, P.K. (1998). Performance of grain legumes in the Indo-Gangetic plain. In *Residual Effects of Legumes in Rice and Wheat Cropping Systems of the Indo-Gangetic Plain*, eds. J.V.D.K. Kumar Rao, C. Johansen, and T.J. Rego. New Delhi and Pantancheru, India: Oxford & IBH Publishing Co. Pvt. Ltd. and ICRISAT, pp. 3-13.

Joshi, P.K., M.C.S. Bantilan, R.L. Shiyani, M. Asokan, and S.C. Sethi. (1998). Chickpea in the hot and dry climate of India-Adoption and impact of improved varieties. In *Assessing Joint Research Impacts: Proceedings of an International*

Workshop on Joint Impact Assessment of NARS/ICRISAT Technologies for the Semi-Arid Tropics, eds. M.C.S. Bantilan and P.K. Joshi, 2-4 December 1996, ICRISAT, Patancheru, India, pp. 94-102.

Karim, Z., M.M.U. Miah, and S.G. Hassain. (1992). Zinc deficiency problems in floodplain agriculture. In *Proceedings of the International Symposium on the Role of Sulphur, Magnesium and Micronutrients in Balanced Plant Nutrition*. ed. S. Portch. Washington, D.C.: The Sulphur Institute. pp. 337-343.

Karki, T.B. (1993). Country Report 9-Nepal. In *Soybean in Asia*, eds. N. Chomchalow, P. Laosuwan, RAPA Publication:1993/6, Lebanon, NH: Science Publishers, Inc., pp. 79-86.

Khatri, B.B., and G.J. Wells. 1999. Effect of grain and forage legumes on the yield of rice and potato-based cropping systems in Nepal. Manila, Phillippines: International Rice Research Institute. International Rice Research Notes 24: 34-35.

Kolar, J.S., H.S. Grewal, and, B. Singh. (1993). Nitrogen substitution and higher productivity of a rice-wheat cropping system through green manuring. *Tropical Agriculture* 70: 301-304.

Kolar, J.S., Mohinder-Singh, K.S. Sandhu, and A.S. Sidhu. (1979). Studies on weed control in mung bean (*Vigna radiata* var. *aureus*). *Journal of Research*, Punjab Agricultural University. 16: 14-18.

Kumar, V. and A.C. Kapoor. (1984). Trace mineral composition of different varieties of cereals and legumes. *The Indian Journal of Nutrition and Dietetics* 21:137-143.

Kurien, V. (2000). India's milk revolution. *Nutrition Foundation of India Bulletin* 21:1-10.

Kushwaha, B.L. (1993). Response of urdbean to phosphorus and zinc application. *Indian Journal of Pulses Research* 6(2): 152-154.

Ladha, J.K., D.K. Kundu, M.G. Angelo-Van Coppenolle, M.B. Peoples, V.R. Carangal, and P.J. Dart. (1996). Legume productivity and soil nitrogen dynamics in lowland rice-based cropping systems. *Soil Science Society of America Journal* 60: 183-192.

Ladha, J.K., I. Watanabe, and S. Saono. (1988). Nitrogen fixation by leguminous green manure and practices for its enhancement in tropical lowland rice. In *Green Manure in Rice Farming*. Int. Rice Res. Inst., Los Banos, Philippines pp. 165-183.

Lambein, F., R. Haque, J.K. Khan, N. Kebede, and Y.H. Kuo. (1994). From soil to brain: Zinc deficiency increases the neurotoxicity of *Lathyrus sativus* and may affect the susceptibility for the motorneurone disease neurolathyrism. *Toxicon* 32(4): 461-466.

Lauren, J.G., J.M. Duxbury, V. Beri, M.A. Razzaque II, M.A. Sattar, S.P. Pandey, S. Bhattarai, R.A. Mann, and J.K. Ladha. (1998). Direct and residual effects from forage and green manure legumes in rice-based cropping systems. In *Residual Effects of Legumes in Rice and Wheat Cropping Systems of the Indo-Gangetic Plain*, eds. J.V.D.K. Kumar Rao, C. Johansen, and T.J. Rego. New Delhi and Pantancheru, India: Oxford & IBH Publishing Co. Pvt. Ltd. and ICRISAT, pp. 55-81.

Leng, R.A. (1991). Improving ruminant production and reducing methane emissions from ruminants by strategic supplementation. United States Environmental Protection Agency, EPA/400/1-91/004.

Lizhi, C. (1988). Green manure cultivation and use for rice in China. In *Green Manure in Rice Farming*. Los Banos, Philippines: International Rice Research Institute, pp. 63-70.

Mahapatra, B.S. and G.L. Sharma (1989). Integrated management of *Sesbania*, *Azolla* and urea nitrogen in lowland rice under a rice-wheat cropping system. *Journal of Agricultural Science* 113: 203-206.

Mahapatra, B.S., K.C. Sharma, and G.L. Sharma. (1987). Effect of bio, organic and chemical N on yield and N uptake in rice and their residual effect on succeeding wheat crop. *Indian Journal of Agronomy* 32: 7-11.

Malik, B.A. (1987). Food legume cultivars released to improve farming systems in Pakistan. In *Food Legume Improvement for Asian Farming Systems: Proceedings of an International Workshop*, eds. E.S. Wallis and D.E. Byth, 1-5 September 1986, Khon Kaen, Thailand. ACIAR Proceedings No. 18. Canberra, Australia: ACIAR, pp. 318.

McWilliam, J.R. and J.L. Dillon (1987). Food legume crop improvement: Progress and constraints. In *Food Legume Improvement for Asian Farming Systems: Proceedings of an International Workshop*, eds. E.S. Wallis and D.E. Byth, 1-5 September 1986, Khon Kaen, Thailand. ACIAR Proceedings No. 18. Canberra, Australia: ACIAR, pp. 22-33.

Meelu, O.P. and R.S. Rekhi. (1981). Fertilizer use in rice based cropping system in northern India. *Fertilizer News* 26: 16-22.

Mehra, R.B., D.B. Raju, and K. Himabindu. (1995). Breeding Work on *Lathyrus sativus* L. at I.A.R.I., New Delhi. In *Lathyrus sativus and Human Lathyrism: Progress and Prospects, Proceedings of Second International Colloquium*. eds. H. Yusef and F. Lambein, 10-12 December 1993. Dhaka, Bangladesh, The University of Dhaka. pp. 127-130.

Miah, A.A. and M.M. Rahman. (1993). Agronomy of lentil in Bangladesh. In *Lentil in South Asia: Proceedings of the Seminar on Lentil in South Asia*, eds. W. Erskine and M.C. Saxena, 11-15 March 1991, New Delhi, India. ICARDA, Aleppo, Syria, pp. 128-138.

Mondal, A.K., S. Pal, M. Biswapat, and L.N. Mandal. (1991). Available boron and molybdenum content in some alluvial acidic soils of North Bengal. *Indian Journal of Agricultural Sciences* 61(7): 502-504.

Moog, F.A. (1986). Forages in integrated food cropping systems. pp. 152-160. In *Forages in Southeast Asian and South Pacific Agriculture*, eds. G.J. Blair, D.A. Ivory and T.R. Evans Australian Centre for International Agricultural Research, Proceedings Series No. 12.

Mosier, A.R., J.M. Duxbury, J.R. Freney, O. Heinemeyer, K. Minami, and D.E. Johnson. (1998). Mitigating agricultural emissions of methane. *Climatic Change* 40: 39-80.

Muehlbauer, F.J. and W.J. Kaiser. eds, (1994). *Expanding the Production and Use of Cool Season Food Legumes–Proceedings of the Second International Food Legume Research Conference on Pea, Lentil, Faba Bean, Chickpea, and Grasspea*, 12-16 April 1992, Cairo, Egypt. Dordrecht, The Netherlands: Kluwer Academic Publishers, 991 p.

Musa, A.M., C. Johansen, J. Kumar, and D. Harris. (1999). Response of chickpea to seed priming in the high Barind tract of Bangladesh. *International Chickpea and Pigeonpea Newsletter* 6: 20-22.

Nagarajah, S., H.U. Neue, and M.C.A. Alberto. (1989). Effect of *Sesbania, Azolla* and rice straw incorporation on the kinetics of NH_4, K, Fe, Mn, Zn and P in some flooded rice soils. *Plant and Soil* 116: 37-48.

Nagaraju, A.P. and Y.H. Yadahalli. (1996). Response of cowpea (*Vigna unguiculata*) to sources and levels of phosphorus and zinc. *Indian Journal of Agronomy* 41(1): 88-90.

NARC. (1998). NARC Research Highlights: 1989/90 to 1994/95. Nepal Agricultural Research Council, Khumaltar, Nepal.

Neupane, R.K. and M.P. Bharati. (1993). Agronomy of lentil in Nepal. In *Lentil in South Asia*: *Proceedings of the Seminar on Lentil in South Asia*, eds. W. Erskine and M.C. Saxena, 11-15 March 1991, New Delhi, India. ICARDA, Aleppo, Syria, pp. 139-145.

Norman, M.J.T., C.J. Pearson, and P.G.E. Searle. (1995). *The Ecology of Tropical Food Crops*, Cambridge, England: Cambridge University Press, pp. 185-192.

Pandey, S.P., C.R. Yadav, K. Sah, S. Pande, and P.K. Joshi. (2000). Legumes in Nepal. In *Legumes in Rice and Wheat Cropping Systems of the Indo-Gangetic Plain–Constraints and Opportunities*. eds. C. Johansen, J.M. Duxbury, S.M. Virmani, C.L.L. Gowda, S. Pande, and P.K. Joshi. Pantancheru, India and Ithaca, NY: ICRISAT and Cornell University. In press.

Pariselle, A., and G. Rinaudo. (1988). Studies of the interactions between *Sesbania rostrata* and *Hirshmanniella oryzae* and rice yields. *Revue de Nematologie* 11: 83-87.

Pasricha, N.P. and G.S. Bahl. (1996). Response to applied potassium for yield and quality in oilseed and pulse crops: A Review. *Fertilizer News* 41(5): 27-37.

Peoples, M.B. and D.F. Herridge. (1990). Nitrogen fixation by legumes in tropical and subtropical agriculture. *Advances in Agronomy* 44: 155-223.

Peoples, M.B. and E.T. Craswell. (1992). Biological nitrogen fixation: Investments, expectations and actual contributions to agriculture. *Plant and Soil* 141: 13-39.

Phillips, R.D. (1993). Starchy legumes in human nutrition, health and culture. *Plant Foods for Human Nutrition* 44: 195-211.

Rahman, A. (1994). *A Study on the Production Potential of Forage Crop in Two Predominant Cropping Systems of Bangladesh*. Ph.D. Thesis Bangladesh Agricultural University, Mymensingh, Bangladesh. pp. 260.

Rahman, M.M. and A.A. Miah. (1988). Mungbean in Bangladesh–problems and prospects. In *Mungbean*: *Proceedings of the Second International Symposium*, 16-20 November 1987, Bangkok, Thailand. Asian Vegetable Research and Development Center (AVRDC), Shanhua, Tainan: AVRDC, pp. 570-579.

Rahman, M.M., M. Quader, and J. Kumar. (1991). Status of khesari breeding and future strategy. In *Advances in Pulses Research in Bangladesh–Proceedings of the Second National Workshop on Pulses*, 6-8 Jun. 1989, Joydebpur, Bangladesh. Bangladesh Agricultural Research Institute, International Development Research Centre and ICRISAT. Pantancheru, India: ICRISAT, pp. 25-28.

Rahman, M.M., M.A. Bakr, M.F. Mia, K.M. Idris, C.L.L. Gowda, J. Kumar, U.K. Deb, M.A. Malek, and A. Sobhan. (2000). Legumes in Bangladesh. In *Legumes in*

Rice and Wheat Cropping Systems of the Indo-Gangetic Plain–Constraints and Opportunities. eds. C. Johansen, J.M. Duxbury, S.M. Virmani, C.L.L. Gowda, S. Pande, and P.K. Joshi. Pantancheru, India and Ithaca, NY: ICRISAT and Cornell University. In press.

Rashid, A., J. Din and M. Bashir. (1999). Phosphorus deficiency diagnosis and fertilization in mungbean grown in rainfed calcareous soils of Pakistan. *Communications in Soil Science and Plant Analysis.* 30: 2045-2056.

Rathore, A.L., S.J. Chipde, and A.R. Pal. (1995). Direct and residual effects of bio-organic and inorganic fertilizers in rice (*Oryza sativa*)-wheat (*Triticum aestivum*) cropping system. *Indian Journal of Agronomy* 40: 14-19.

Raychaudhuri, S.P. (1960). Use of artificial fertilizers in India. *Indian Council of Agriculture Review Series* No. 32. Indian Council of Agricultural Research, New Delhi.

Reddy, K.C., A.R. Soffes, G.M. Prine, and R.A. Dunn. (1986). Tropical legumes for green manure. II. Nematode populations and their effects on succeeding crop yields. *Agronomy Journal* 78: 5-10.

Regmi, K.R. (1990). *Trainer's Manual No. 9 Grain Legumes.* Department of Agriculture, Agriculture Training and Manpower Development Programme, Manpower Development Agriculture Project, Kathmandu, Nepal.

Rekhi, R.S. and M.S. Bajwa. (1993). Effect of green manure on the yield, N uptake and floodwater properties of a flooded rice, wheat rotation receiving ^{15}N urea on a highly permeable soil. *Fertilizer Research* 34: 15-22.

Rekhi, R.S. and O.P. Meelu. (1983). Effect of complementary use of mung straw and inorganic fertilizer nitrogen on nitrogen availability and yield of rice. *Oryza (India)* 20: 125-129.

Rekib, A. (1980). A net profit of Rs 5,600 by cowpea dairy farming. *Indian Farming* 30(2): 21.

Rukonuddin, A., M. Kabirullah, and S.A. Khan. (1994). Preparation of poultry feeds from indigenous raw materials, feed for growers. *Bangladesh Journal of Scientific and Industrial Research* 29(3): 11-17.

Sakal, R. (1993). Micronutrient status of Bihar soils and response of crops to micronutrient application. In *Proceedings of the Workshop on Micronutrients.* ed. M.H. Ali. 22-23 January, 1992, Bhubaneswar, India. Indo-British Fertilizer Education Project, Hindustan Fertilizer Corporation, pp. 199-213.

Sakal, R., R.B. Sinha, A.P. Singh, and N.S. Bhogal. (1998). Response of some rabi pulses to boron, zinc and sulphur application in farmer's field. *Fertilizer News* 43(11): 37, 39-40.

Saraf, C.S., O.P. Rupela, D.M. Hegde, R.L. Yadav, B.G. Shivakumar, S. Bhattarai, M.A. Razzaque II, and M.A. Sattar. (1998). Biological nitrogen fixation and residual effects of winter grain legumes in rice and wheat cropping systems of the Indo-Gangetic Plain. In *Residual Effects of Legumes in Rice and Wheat Cropping Systems of the Indo-Gangetic Plain*, eds. J.V.D.K. Kumar Rao, C. Johansen, and T.J. Rego. New Delhi and Pantancheru, India: Oxford & IBH Publishing Co. Pvt. Ltd. and ICRISAT, pp. 14-30.

Sarker, A., M. Rahman, and W. Zaman. (1992). Utfala: A lentil variety for Bangladesh. *LENS Newsletter* 19(1): 14-15.

Sarker, A., W. Erskine, M.S. Hassan, M.A. Afzal, and A.N.M.M. Murshed. (1999). Registration of 'Barimasur-4' lentil. *Crop Science* 39(3): 876.

Saxena, N.P. and C. Johansen. (1990). Realized yield potential in chickpea and physiological considerations for further genetic improvement. In *Realized Yield Potential in Chickpea and Physiological Considerations for Further Genetic Improvement: Proceedings of the International Congress of Plant Physiology*, eds. S.K. Sinha, P.V. Sane, S.C. Bhargava, P.K. Agrawal, 15-20 February 1988. New Delhi, India: Society for Plant Physiology and Biochemistry. Vol. 1: 279-288.

Schiere, J.B., M.N.M. Ibrahim, and A. de Rond. (1985). Supplementation of urea-treated rice straw. In *Relevance of Crop Residues as Animal Feeds in Developing Countries, Proceedings of an International Workshop*, eds. M. Wanapat and C. Devendra, November 29-December 2, 1984. Khon Kaen, Thailand, Bangkok, Thailand: Funny Press.

Schulz, S. (1997). Performance and residual effects of leguminous crops in rice-based cropping systems of the middle mountains of Nepal. Ph.D Thesis, Department of Agriculture, The Reading University, UK.

Seshadri, S. (1997). Nutritional anaemia in South Asia. In *Malnutrition in South Asia: A Regional Profile*, ed. S. Gillespie, ROSA Publication No. 5, UNICEF-South Asia, pp. 1-189.

Shaikh, M.A.Q., M.A. Majid, and B.P. Lahiri. (1995). Mutant deriviative low neuro-toxin lines of grass pea. In *Lathyrus sativus and Human Lathyrism: Progress and Prospects, Proceedings of Second International Colloquium*. eds. H. Yusef and F. Lambein, 10-12 December 1993. Dhaka, Bangladesh, The University of Dhaka. pp. 165-172.

Shaikh, M.A.Q., Z.U. Ahmed, and R.N. Oram. (1988). Winter and summer mung-bean breeding for improved disease resistance and yield. In *Mungbean: Proceedings of the Second International Symposium*, 16-20 November 1987, Bangkok, Thailand. Asian Vegetable Research and Development Center (AVRDC), Shanhua, Tainan: AVRDC, pp. 124-129.

Sharma, A.R. and B.N. Mittra. (1988). Effect of green manuring and mineral fertilizer on growth and yield of crops in rice-based cropping on lateritic soil. *Journal of Agricultural Science* 110: 605-608.

Sharma, C.M., R.S. Minhas, and S.S. Masand. (1988). Molybdenum in surface soils and its vertical distribution in profiles of some acid soils. *Journal of Indian Society of Soil Science* 36(2): 252-256.

Sharma, R.S., and C.L. Thakur. (1993). Integrated and economically viable weed management in soybean (*Glycine max*)-wheat (*Triticum aestivum*) crop sequence. *Indian Journal of Agricultural Sciences* 63: 556-560.

Sharma, S.K., R.N. Verma, and S. Chauhan. (1996). Effect of NPK doses on yield and severity of three major diseases of soybean at medium altitude of East Khasi Hills (Meghalaya). *Journal of Hill Research* 9(2): 279-295.

Sharma, S.N., R. Prasad, and S. Singh. (1995). The role of mungbean residues and *Sesbania aculeata* green manure in the nitrogen economy of rice-wheat cropping system. *Plant and Soil* 172: 123-129.

Singh, B.D. and B.P. Singh. (1998). Effect of weed management practices and phosphorus level on weed infestation, nodulation and yield of chickpea and mustard intercropping system. *Indian Journal of Weed Science* 30(3/4): 124-128.

Singh, H.P., A. Singh, and S. Singh. (1995). Nodulation and growth of soyabeans as influenced by rock phosphate, *Bradyrhizobium*, mycorrhiza and phosphate-dissolving microbes. *Bhartiya Krishi Anusandhan Patrika* 10(3): 117-124.

Singh, K., G. Habib, M.M. Siddiqui, and M.N.M. Ibrahim. (1997). Dynamics of feed resources in mixed farming systems of South Asia. In *Crop Residues in Sustainable Mixed Crop/Livestock Farming Systems*, ed. C. Renard, CAB International, NY, NY, pp. 113-130.

Singh, L. and Y.C. Srivastava. (1988). Blackgram. In *Pulse Crops*, eds. B. Baldev, S. Ramanujam, and H.K. Jain, New Delhi, India: Oxford and IBH Publishing Co. Pvt. Ltd., pp. 189-198.

Singh, M.V. (1999). Current status of micro and secondary nutrients deficiencies and crop response in different agro-ecological regions: Experiences of all india co-ordinated research project on micro and secondary nutrients and pollutant elements in soils and plants. *Fertilizer News* 44(4): 63-82.

Singh, P. (1988). *Indian rangelands status and improvement*. Plenary Address at the Third International Rangeland Congress, 7-11 November 1988, New Delhi, India.

Singh, R.C., M. Singh, R. Kumar, and D.P.S. Tomar. (1994). Response of chickpea (*Cicer arietinum*) genotypes to row spacing and fertility under rainfed conditions. *Indian Journal of Agronomy* 39(4): 569-572.

Singh, R.P., and H. Chand. (1996). Effect of host nutrition on Ascochyta blight disease in chickpea. *Indian Journal of Pulses Research* 9(1): 85-86.

Singh, S.P. (1988). Pigeonpea. In *Pulse Crops*, eds. B. Baldev, S. Ramanujam, and H.K. Jain, New Delhi, India: Oxford and IBH Publishing Co. Pvt. Ltd., pp. 132-160.

Singh, V.P., A. Chhabra, and R.P.S. Kharb. (1988). Production and utilization of mungbean in India. In *Mungbean: Proceedings of the Second International Symposium*. 16-20 November 1987, Bangkok, Thailand. Asian Vegetable Research and Development Center (AVRDC), Shanhua, Tainan: AVRDC, pp. 486-497.

So, H.B. and T. Woodhead. (1987). Alleviation of soil physical limits to productivity of legumes in Asia. In *Food Legume Improvement for Asian Farming Systems: Proceedings of an International Workshop*, eds. E.S. Wallis and D.E. Byth, 1-5 September 1986, Khon Kaen, Thailand. ACIAR Proceedings No. 18. Canberra, Australia: ACIAR, pp. 112-127.

Soil Management Collaborative Support Research Program (SM-CRSP). (2000). *Sustainability of Post-Green Revolution Agriculture: The Rice-Wheat Cropping System of South Asia*, Cornell University Annual Report to the US Agency for International Development Feb. 10, 1999-Feb. 11, 2000.

Solanki, I.S., R.S. Waldia, V.P. Singh, B.P.S. Malik, and P.S. Kakkar. (1993). Sapna–a high yielding lentil variety for crop rotation. *Indian Farming* 43(8): 9-10.

Srivastava, S.P., M. Joshi, C. Johansen, and T.J. Rego. (1999). Boron deficiency of lentil in Nepal. *LENS Newsletter* 26(1&2): 22-24.

Summerfield, R.J. and E.H. Roberts. eds. (1985). *Grain Legume Crops*. London, England: William Collins Sons & Co. Ltd., 859 p.

Thind, H.S. and Chahal, D.S. (1983). Iron equilibria in submerged soils as influenced by green manuring and iron application. *Journal of Agricultural Science* 101: 207-221.

Thomas, A., G.L. Soni, and R. Singh (1987). Role of dietary fiber from bengal gram as hypocholesterolaemic agent–effect on erythrocyte membrane-bound enzymes and hematological parameters. *Human Nutrition. Food Sciences and Nutrition* 41F (3-4). 193-201.

Tickoo, J.L. and H.K. Jain. (1988). Mungbean. In *Pulse Crops*, eds. B. Baldev, S. Ramanujam, and H.K. Jain, New Delhi, India: Oxford and IBH Publishing Co. Pvt. Ltd., pp. 161-188.

Tiwari, K.N., S.P. Tiwari, and A.N. Pathak. (1980a). Studies on green manuring of rice in double cropping system in a partially reclaimed saline sodic soil. *Indian Journal of Agronomy* 25: 136-145.

Tiwari, K.N., A.N. Pathak, and H. Ram. (1980b). Green-manuring in combination with fertilizer nitrogen on rice under double cropping system in alluvial soil India. *Journal of the Indian Society of Soil Science* 28: 162-169.

Tufail, M., M. Ahmad, and A. Ali. (1995). Masoor-93: An ideal combination of characters for Punjab province. *Lens Newsletter* 22(1/2): 50-52.

U.S. Department of Agriculture, Agricultural Research Service. 1999. *USDA Nutrient Database for Standard Reference, Release 13. Nutrient Data*. Laboratory Home Page, *http://www.nal.usda.gov/fnic/foodcomp*

Vaidya, Y. (1990). *Young child feeding practices and feasibility of local weaning food production*. Central Food Research Food Laboratory, Minister of Local Development, Kathmandu, Nepal.

Wallis, E.S. and D.E. Byth eds. (1987). *Food Legume Improvement for Asian Farming Systems: Proceedings of an International Workshop*, 1-5 September 1986, Khon Kaen, Thailand. ACIAR Proceedings No. 18. Canberra, Australia: ACIAR. 341 p.

Wallis, E.S., K.B. Saxena, C. Brauns, D.E. Byth, and P.C. Whiteman. (1983). Short-season pigeonpeas–a cropping system with high yield potential. *International Pigeonpea Newsletter* 2: 5-8.

Yaacob, O. and G.J. Blair. (1981). Effect of legume cropping and organic matter accumulation on the infiltration rate and structural stability of a granite soil under a simulated tropical environment. *Plant and Soil* 60: 11-20.

Yadav, R.L., B.S. Dwivedi, K.S. Gangwar, and K. Prasad. (1998). Overview and prospects for enhancing residual benefits of legumes in rice and wheat cropping systems in India. In *Residual Effects of Legumes in Rice and Wheat Cropping Systems of the Indo-Gangetic Plain*, eds. J.V.D.K. Kumar Rao, C. Johansen, and T.J. Rego. New Delhi and Pantancheru, India: Oxford & IBH Publishing Co. Pvt. Ltd. and ICRISAT, pp. 207-225.

Yadvinder-Singh, B. Singh, M.S. Maskina, and O.P. Meelu. (1988). Effect of organic manures, crop residues and green manure (*Sesbania aculeata*) on nitrogen and phosphorus transformations in a sandy loam at field capacity and under water-logged conditions. *Biology and Fertility of Soils* 6: 183-187.

Yadvinder-Singh, C.S. Khind, and B. Singh. (1991). Efficient management of leguminous green manures in wetland rice. *Advances in Agronomy* 45: 135-189.

Zahid, M.A., R.A. Mann, and Z. Shah. (1998). Overview and prospects for enhancing residual benefits of legumes in rice and wheat cropping systems in Pakistan. In *Residual Effects of Legumes in Rice and Wheat Cropping Systems of the Indo-Gangetic Plain*, eds. J.V.D.K. Kumar Rao, C. Johansen, and T.J. Rego. New Delhi and Pantancheru, India: Oxford & IBH Publishing Co. Pvt. Ltd. and ICRISAT, pp. 190-206.

Zohary, D. and M. Hopf. (1973). Domestication of pulses in the old world. *Science* 182: 887-894.

Policy Re-Directions
for Sustainable Resource Use:
The Rice-Wheat Cropping System
of the Indo-Gangetic Plains

P. L. Pingali
M. Shah

SUMMARY. The last three decades or more have witnessed the phenomenal growth of cereal crop productivity in the developing world, particularly for rice and wheat in Asia. High levels of investments in research and infrastructure development, especially irrigation infrastructure, resulted in the rapid intensification of the lowlands. Consequently, the irrigated and the high rainfall lowland environments became the primary source of food supply for Asia's escalating population. The emergence of the rice-wheat system in South Asia as the most important source of food supply, is a testament to the success of the Green revolution in wheat and rice.

Recent signs indicate, however, a slow down in productivity growth of the primary cereals, rice and wheat, especially in the intensively cultivated lowlands of Asia, and particularly in the intensively cultivated rice-wheat zones of South Asia. Slackening of infrastructure and research investments and reduced policy support partly explain the sluggish growth. This paper argues that in addition to the above factors,

P. L. Pingali is Director, Economics Program, and M. Shah is Research Associate, Centro Internacional de Mejoramiento de Maiz y Trigo (CIMMYT), Mexico.

Address correspondence to: P. L. Pingali, Director, Economics Program, Lisboa 27, Apartado Postal 6-641, 06600, Mexico, D.F., Mexico (E-mail: p.pingali@cgiar.org).

[Haworth co-indexing entry note]: "Policy Re-Directions for Sustainable Resource Use: The Rice-Wheat Cropping System of the Indo-Gangetic Plains." Pingali, P L., and M. Shah. Co-published simultaneously in *Journal of Crop Production* (Food Products Press, an imprint of The Haworth Press, Inc.) Vol. 3, No. 2(#6), 2001, pp. 103-118; and: *The Rice-Wheat Cropping System of South Asia: Trends, Constraints, Productivity and Policy* (ed: Palit K. Kataki) Food Products Press, an imprint of The Haworth Press, Inc., 2001, pp. 103-118. Single or multiple copies of this article are available for a fee from The Haworth Document Delivery Service [1-800-342-9678, 9:00 a.m. - 5:00 p.m. (EST). E-mail address: getinfo@haworthpressinc.com].

degradation of the lowland resource base due to intensive use, over the long term, also contributes to declining productivity growth rates. Intensification *per se* is not the root cause of lowland resource base degradation, but rather the policy environment that encouraged inappropriate land use and injudicious input use, especially water and chemical fertilizers.

Trade policies, output price policies as well as input subsidies have all contributed to the unsustainable use of the lowlands. The dual goals of food self-sufficiency and sustainable resource management are often mutually incompatible. Policies designed for achieving food self-sufficiency tend to undervalue goods not traded internationally, especially land and labor resources. As a result, food self-sufficiency in countries with an exhausted land frontier, particularly the countries of South Asia, came at a high ecological and environmental cost. Appropriate policy reform, both at the macro as well as at the sector level will go a long way towards arresting and possibly reversing the current degradation trends.

This paper provides an extensive review of the existing evidence on intensification induced degradation in the rice-wheat systems of the Indo-Gangetic plains of South Asia. Policy re-directions and corrections that can contribute to arresting and/or reversing the current degradation trends are recommended. *[Article copies available for a fee from The Haworth Document Delivery Service: 1-800-342-9678. E-mail address: <getinfo@haworthpressinc.com> Website: <http://www.HaworthPress.com>* © 2001 by The Haworth Press, Inc. All rights reserved.]

KEYWORDS. Policy, rice, South Asia, sustainable, wheat

INTRODUCTION

In 1950-52 rice production for all of South Asia was only 47 million tons and by 1996-98 it was 161.5 million tons with India being the largest producer at 123 million tons. We note a similar story with wheat production in South Asia. Production increased from 9.8 million tons in 1950-52 to 85.8 million tons in 1996-98; with India being the largest producer at around 66 million tons. Chapter 1 of this publication (Kataki, Hobbs & Adhikary, 2000) provides an account of wheat and rice production from 1958-1998 for South Asia. It illustrates the quantum gains in production that have occurred in the region over time. Between 1958 and 1998 the growth rates of rice and wheat production in South Asia (2.5% and 5.2% per annum, respectively) exceeded the rate of growth in population (2.22% per annum), indicating an increase over time in the per capita availability of the two cereals.

The dramatic increase in production came from the intensification of land use, and to yield growth, the former attributable to investments in irrigation infrastructure and the latter to the adoption of modern seed-fertilizer technologies. For example, by 1979, almost 100% of the wheat planted in the Indian Punjab, the largest wheat producer in India, were to HYVs. The average wheat yield in the Indian Punjab doubled from 2 MT/ha in 1970-71 to over 4 MT/ha in 1993-94. The area under the rice-wheat cropping systems in India grew at 3.2% annually from 1959-62 to 1986-89.

While the production of wheat and rice has been increasing at such a rapid rate, real food prices have been declining steadily since mid 1960s. Figure 1 provides an example for wheat in India. A similar decline in the real price of rice has been observed in India and across Asia (Pingali, Hossain and Gerpacio, 1997). The temporal decline in basic food prices has been especially beneficial to urban consumers, particularly the urban poor, as well as the rural poor who tend to be net purchasers of food.

The major factors that contributed to the initial success of the Green Revolution, and to the emergence of the rice-wheat system as an

FIGURE 1. Wheat production and real prices for India

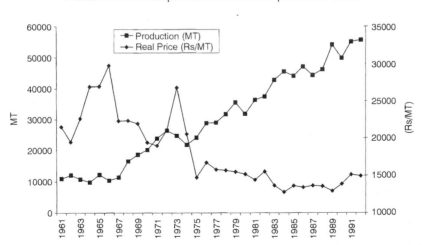

Source: FAOSTAT Online database; IMF, International Financial Statistics and Government of India.

important source of food supply, were: introduction of the high yielding, semi-dwarf varieties of rice and wheat (modern germplasm); infrastructure investments, especially irrigation systems; political commitment and policy support. The latter two were as important as the first two in the rapid dissemination and adoption of modern technologies and the rapid growth in food production. The commitment to achieving food self sufficiency was the driving political force that made the Green Revolution happen in South Asia. Micro- and Macroeconomic policies that promoted rapid productivity growth through the adoption of modern wheat and rice technologies were established in the mid-1960s.

In the early years, input price subsidies and output price supports were essential as they helped stimulate farmers to adopt new technologies. Unpriced irrigation water, cheap fertilizers, subsidized power supply and low interest farm credit were some of the crucial supports provided by South Asian governments that made intensive rice-wheat production profitable. However, prolonging the policies of input price subsidies into the post-green revolution period resulted in a distortion of farm level incentives for efficient input use and lead to much of the degradation observed today.

While micro policies have played an important role in leading South Asian farmers to unsustainable agricultural practices, macro-policy scenarios have been just as important. Since food self-sufficiency was the motivating factor for many of the policy measures during the 1970s and 1980s, macroeconomic policies protected cereal prices through import restrictions and tariffs. Domestic prices were kept artificially high and excessive productive resources were devoted to the production of rice and wheat. These were 'safe' crops that would get farmers assured prices at subsidized input prices. Thus there was no real incentive for farmers to diversify from the rice-wheat rotation. The impact that past micro and macro economic policies have had on the rice-wheat resource base utilization and degradation are discussed below.

ECOLOGICAL CONSEQUENCES OF INTENSIFICATION

Intensive rice-wheat rotation on the lowlands results in the following changes in production systems: (i) seasonal wet and dry crop cycles over the long term; (ii) increased reliance on irrigation and inorganic fertilizers; (iii) asymmetry of planting schedules; and (iv) greater uniformity in the varieties cultivated. Over the long term, these changes impose significant environmental costs due to negative biophysical impacts.

The most common environmental consequences of lowland intensification are: (i) buildup of salinity and water logging; (ii) depletion/pollution of (ground) water resources; (iii) formation of a hardpan (subsoil compaction); (iv) changes in soil nutrient status, nutrient deficiencies and increased incidence of soil toxicity; and (v) increased pest buildup, pest-related yield losses and associated consequences of increased and injudicious pesticide use. A brief description of each of these problems and the possibilities for reversing them are discussed below. At the farm level, long-term changes in the biophysical environment are manifested in terms of declining total factor productivity, profitability, and input efficiencies. Many of the degradation problems mentioned above were policy induced and the result of inappropriate and inefficient land, water and input use practices.

Salinity and Water Logging

Intensive use of irrigation water in areas with poor drainage can lead to a rise in the water table due to the continual recharge of groundwater. In the semi-arid and arid zones this leads to salinity buildup, while in the humid zone to water logging. An excess of evapotranspiration over rainfall causing a net upward movement of water through capillary action and the concentration of salts on the soil surface induces salinity. The groundwater itself need not be saline for salinity to build up; it can occur due to the long-term evaporation of continuously recharged water of low salt content (Moormann and van Breeman, 1978).

Excessive irrigation, poor drainage especially, seepage can induce salinity problems from unlined canals. Poor irrigation system design and management are primary factors leading to salinity problems. Irrigation water provided free or at a very low cost to the farmer tends to aggravate the problem. For instance, in Pakistan's Sindh Province, large areas became saline after the introduction of extensive irrigation. This led to a rise of the water table from a depth of 20-30 m to 1-2 m within 20 years (Moormann and Van Breeman, 1978); other examples from South Asia can be found in Chambers (1988), Abrol (1987), Dogra (1986) and Harrington et al. (1992). Dogra (1986) estimates that in India, nearly 4.5 million hectares are affected by salinization and a further 6 million by water logging. In the short term salinity buildup leads to reduced yields while in the long term it can lead to abandoning of crop lands (Samad et al., 1992; Postel 1989; Mustafa 1991).

In higher rainfall areas, such as in East India, induced salinity buildup is not as much a problem because the rain flushes out the accumulated salts. However, excessive water use and poor drainage cause problems of water logging in this zone. Waterlogged fields have lower productivity levels because of lower decomposition rates of organic matter, lower nitrogen availability, and accumulation of soil toxins. In the case of wheat, low plant populations in some areas can be attributed to water logging, especially water logging occurring early in the growing season during germination and emergence stages of wheat. Hobbs et al. (1996) report for the Nepal Terai that water logging reduced yields by half a ton per hectare.

Groundwater Depletion

Development of groundwater resources has been a significant driving force for agricultural intensification in many parts of Asia. The massive expansion of private sector tubewell irrigation in Bangladesh, India, and Pakistan is the most successful example of private sector irrigation development in Asia. A "groundwater revolution" in Bangladesh beginning in the 1980s was a key stimulant to rapid agricultural growth in the 1980s and early 1990s. Nearly 1.5 million hectares of land were newly irrigated after 1980, in significant part from installation of shallow tubewells spurred by deregulation of tubewell imports (Rogers et al., 1994).

However, just as excess use of un-priced irrigation water can lead to rising water tables and salinization, it can also lead to falling water tables, in tube well irrigated areas, with negative environmental and productivity consequences. The problem of overdrafting of groundwater often occurs because individual pump irrigators have no incentive to optimize long-run extraction rates, since water left in the ground can be captured by other irrigators or potential future irrigators. Groundwater is depleted when pumping rates exceed the rate of natural recharge of the aquifer. While mining of both renewable and non-renewable water resources can be an optimal economic strategy, it is clear that groundwater overdrafting is excessive in many intensive agricultural areas in Asia. The problem of ground water overdrafting is exacerbated when electricity for tubewell operations is subsidized.

Government intervention to prevent depletion of groundwater in the developing world has proven difficult to implement, subject to corruption, and in many cases very costly. The most successful tubewell development has been through small-scale private investment, which is widely dispersed and difficult to monitor. Only when private tubewell imports and markets were deregulated did the small-scale tubewell revolution take off in Bangladesh. An attempt at re-regulation through restrictions on well-siting slowed growth in tubewell adoption during 1985-1987 (Rogers et al., 1994). India has been ineffective at implementing licensing laws at the state level, where ownership of all water resources resides. Pakistan has no legal system for licensing groundwater withdrawals, and limited attempts to give ownership of underlying aquifers to municipalities have been challenged in the courts (Pingali and Rosegrant, 1998).

Changes in Soil Nutrient Status

The most commonly observed effect of intensive rice-wheat systems is a decline in the partial factor productivity of nitrogen fertilizer (Hobbs and Morris, 1996). Recent work at IRRI (Cassman et al., 1994) indicates that the declining partial factor productivity of nitrogen in rice monoculture systems is due to a decline in the nitrogen supplying capacity of intensively cultivated wetland soils. Rice-wheat systems could be facing a similar phenomenon. Fertilized rice and wheat obtain 50-80% of their nitrogen requirement from the soil; unfertilized rice obtains an even larger portion, mainly through the mineralization of organic matter (De Datta, 1981). The soil's capacity to provide nitrogen to the plant declines with continuous (two to three crops per year) flooded rice cultivation systems. Declining soil N supply results in declining factor productivity of chemical nitrogen, since soil N is a natural substitute for chemical nitrogen. Cassman and Pingali (1993) estimate the magnitude of yields foregone due to declining soil nitrogen supply. Using long-term experiment data from the IRRI farm, Cassman and Pingali estimate the decline in yields to be around 30 percent, over a 20-year period, at all nitrogen levels.

In addition to nitrogen, phosphorus and potassium are the two other macronutrients demanded by the rice and wheat plants. Phosphorus and potassium deficiencies are becoming widespread across Asia in areas not previously considered to be deficient. These deficiencies are directly related to the increase in cropping intensity and the predominance of year-round irrigated production systems. For example in China, it is estimated that about two-thirds of agricultural land is now deficient in phosphorus, while in India, nearly one-half of the districts have been classified as low in available phosphorus (Stone, 1986; Tandon, 1987; Desai and Gandhi, 1989). Desai and Gandhi note that this is due to the emphasis on nitrogen rather than a balanced application of all macronutrients required for sustaining soil fertility. The result of unbalanced application of fertilizers has been a decline in the efficiency of fertilizer use over time (Desai and Gandhi, 1989; Stone, 1986; Ahmed, 1985).

Micronutrient Deficiencies and Soil Toxicity

Perennial flooding of ricelands and continuous rice monoculture as well as the rice-rice-wheat rotation lead to increased incidence of micronutrient deficiencies and soil toxicity. Zinc deficiency and iron toxicity are the ones most commonly observed in the tropics. Water logging and salinity buildup, often caused by poor water pricing and water management practices, aggravate these problems. In Asia, zinc deficiency is regarded as a major limiting factor for wetland rice on about 2 million hectares (Ponnamperuma, 1974). In the rice-wheat zone also, zinc deficiency ranks number one in importance among the micronutrient deficiencies (Ivan Ortiz Monasterio, personal communication). These are mainly soils of low zinc content. Soils that are not initially of low zinc content also show signs of induced zinc deficiency due to perennial water saturated conditions and continuous cropping. Drainage, even if temporary, helps alleviate this deficiency by increasing zinc availability (Lopes, 1980; Moormann and van Breeman, 1978).

Most irrigated lowlands do not start off with any soil toxicity; however, toxicity may build up in some soils due to continuous flooding, increased reliance on poor quality irrigation water and impeded drainage, especially on soils where a hardpan is formed due to alternating wet and dry cycles. Iron toxicity is the most commonly observed soil toxicity due to intensive irrigated crop cultivation.

Once diagnosed at the farm level, micronutrient deficiencies are relatively straightforward to correct. Zinc deficiencies can be corrected by adding zinc, for instance. Diagnosis is itself not easy though;

quite often micronutrient deficiencies are misdiagnosed as pest-related damage. In the case of soil toxicity, farm level diagnosis is equally complicated and corrective actions are not as straightforward. In both cases, however, the problem ought to be attacked at the cause rather than the cure stage. Periodic breaks in rice monoculture systems or rice-rice-wheat systems (two crops of rice followed by a crop of wheat) and improved water use efficiency goes a long way towards reducing the incidence and magnitude of the above problems.

Long-Term Changes in Soil Physical Characteristics

Seasonal cycles of puddling (wet tillage) and drying, over the long term, led to the formation of hardpans in paddy soils. The hardpan refers to compacted subsoil that is 5-10 cm thick at depths of 10-40 cm from the soil surface. Compared to the surface soil, a plow pan has higher bulk density and less medium to large-sized pores. Their permeability is generally lower than that of the overlying and deeper horizons. A striking example of the problem of hardpans is seen in the rice-wheat cropping system in South Asia. The poor establishment of wheat following puddled rice affects the productivity of the wheat crop. If the hardpan is broken through deep tillage and soil structures are improved through the incorporation of organic matter, it affects the productivity of the subsequent rice crop by reducing water-holding capacity of the soil. Thus, intensification has reduced the flexibility of dry season crop choice by changing the soil physical structure.

Increasing Losses Due to Pests

The use of purchased inputs for plant protection was unimportant for cereals prior to the mass introduction of modern varieties. Farmers had traditionally relied on host plant resistance, natural enemies, cultural methods, and mechanical methods such as hand weeding. Agricultural intensification in general and continuous cropping of cereals in particular has increased the incidence of weed, insect and disease problems (Pingali and Gerpacio, 1997; Hobbs and Morris, 1996).

In the case of rice, relatively minor pests–leaf folder, caseworm, armyworm and cutworm–started to cause noticeable losses in farmers' fields as area planted to modern varieties increased. Hence the rapid increase in insecticide use in intensive rice monoculture systems (Rola and Pingali, 1993). In the case of wheat, insecticide use is not very prevalent and fungicide use has been largely controlled by the development of varieties with resistance to major disease pressures. However, some diseases, such as *Helminthesporium sativum* (spot blotch) are on the rise in intensive wheat production zones, as well as the rice-wheat zone. Soil borne diseases are also becoming an increasingly important factor in constraining yield growth in the rice-wheat areas of the Indo-Gangetic plains. On the other hand, the incidence of Karnal Bunt, a very important disease problem in wheat has been reduced with the advent of rice-wheat system, because the soil saturated conditions under rice are unfavorable to the disease build up over crop cycles.

Insect and disease problems that have emerged have been exacerbated by crop management and pesticide use practices. Injudicious and indiscriminate pesticide application is related to policies that have made these chemicals easily and cheaply accessible. Heong, Aquino and Barrion (1992) argue that prophylactic pesticide application has led to the disruption of the pest-predator balance and a resurgence of pest populations later in the crop season. Rola and Pingali (1993) have argued that pesticide use has been promoted by policy makers' misperceptions of pests and pest damage. Policy makers commonly perceive that modern variety use necessarily lead to increased pest-related crop losses and that modern cereal production is therefore not possible without high levels of chemical pest control.

Ecologically safe methods of weed management continue to be a major concern for the rice-wheat system. *Phalaris minor* became the major weed problem with the advent of the rice-wheat cropping system in South Asia (Hobbs and Morris, 1996). Homogeneity of cropping patterns across large areas contributes to the rapid build up and spread of *Phalaris*. Breaking up the cropping pattern reduces the weed build up and herbicide resistance problems; however, cropping pattern choices are made on economic grounds rather than sustainability grounds.

The widespread availability of insect and disease resistant varieties for the major cereals has reduced the productivity benefits and the profitability of applying insecticides and fungicides. See Pingali and Gerpacio (1997) for a current review of the impact of host plant resistance for the major cereals and Rola and Pingali (1993) for rice specific evidence. Even where resistant varieties are used one could anticipate pest problems due to a narrowing of genetic diversity on farmers' fields. When many farmers, in the same area, choose to grow the same high yielding variety or ones with similar resistance genes, there is a lower level of genetic diversity than would most effectively protect against the emergence and spread of new disease strains (Heisey et al., 1997). However, increasing diversity on farmers' fields is not a simple proposition, the socially optimum level of diversity might differ quite substantially from the private optimum due to potential yield trade-off and the cost of frequent varietal replacement. See Heisey et al. (1997) for an excellent case study of wheat rust management through enhanced genetic diversity in the Punjab of Pakistan.

POLICIES FOR REVERSING THE CURRENT DEGRADATION TRENDS

Meeting future food requirements in South Asia requires sustained productivity growth from the Indo-Gangetic plains. Continued high levels of investments in research and infrastructure improvement, as well as institutional and policy reforms are necessary to meet the above goal.

While resource base degradation is increasingly observed in the rice-wheat belt, intensification per se is not the root cause of environmental and ecological damage. Severe environmental degradation in intensified agriculture occurs mainly when incentives are incorrect due to bad policy or a lack of knowledge of the underlying processes of degradation.

Government intervention in the cereal market, especially through output price support and input subsidies, provided farmers incentives for increasing cereal productivity. In addition to highly subsidized irrigation water, South Asian farmers benefited from 'cheap' fertilizers, pesticides and credit. The net result was that rice monoculture systems as well as rice-wheat systems were extremely profitable through the decades of the 1970s and the 1980s, despite a long-term decline in the real world rice and wheat prices through this period.

Input subsidies that keep input prices low directly affect crop management practices at the farm level; they reduce farmer incentives for improving input use efficiency, that often requires farmer investment in learning about the technology and how best to use it. As Asian countries liberalize their agricultural sectors and move away from the single-minded pursuit of food self sufficiency one can expect positive resource base and environmental benefits.

Many of the degradation problems observed in the intensively culti-vated rice-wheat lands are not irreversible, appropriate policies will provide farmers the incentives to invest in more sustainable land and crop management practices. Techniques for improving fertilizer use efficiency, for example, are available but will only be viable at the farm level when fertilizer subsidies are removed. The same is the case with the adoption of zero-tillage, Integrated Pest Management tech-niques, or more judicious water management. Table 1 provides details on policy interventions that contribute to resource base sustainability.

In order to create incentives for efficient and more environmentally friendly water allocation, water subsidies (and power subsidies for operation of tube wells) should be phased-out, with more realistic water charges in all sectors. In the longer term, markets in tradable water rights should be established where feasible. Establishment of secure water rights for water users is an important foundation for the establishment of economic incentives for efficient water allocation. Moreover, responsibility for irrigation water management should be devolved where possible to autonomous local institutions with user representation and/or joint ownership. Full financial responsibility should be granted, including right to charge for water and services (Pingali and Rosegrant, 1998).

To complement the approaches to crop management improvement in reducing fertilizer-related degradation problems, fertilizer subsidies ought to be removed to eliminate the incentive for unbalanced and excessive use. The financial costs of fertilizer subsidies to government treasuries are high. The true economic costs can be even greater, as subsidies soak up funds that could be used for alternative investments. The reduction and eventual removal of fertilizer price subsidies can substantially improve the efficiency of fertilizer use.

Non-price policies for fertility management are also important, in-cluding location-specific research on soil fertility constraints and agronomic practices, improvement in extension services, development of improved fertilizer supply and distribution systems, and develop-ment of physical and institutional infrastructure (Desai, 1986; Desai, 1988).

For various environmental and human health reasons (see Pingali

TABLE 1. Reversing the Ecological/Environmental Degradation Problems.

Resource Base Degradation Problem	Possible/Probable Causes	Farm Level Indicators of Resource Degradation	Economic Impact	Possible Technology Intervention	Policy Change
Build up of salinity/ Waterlogging	• Poor design of irrigation systems • Intensive use of irrigation water	• Reduced yields and/or reduced factor productivities • Reduced cropping intensities • Abandoned paddy lands in the extreme	Declining trends in TTP	• Improved irrigation system and design • Increased water use efficiency	• Pricing irrigation water at its 'true' cost
Hardpan (sub-soil compaction)	• Increased frequency of puddling (wet tillage)	• Reduced flexibility non-rice crop the dry season		• Improved farm level drainage systems • Increased water use efficiency	• Pricing irrigation water at its 'true' cost
Changes in soil nitrogen supplying capacity	• Changes in organic matter quantity and quality • Long-term flooding/water saturation of paddy soils	• Declining efficiency of nitrogen fertilizer use • Reduced yields and/or reduced factor productivities	Declining profitability of rice and wheat cultivation Increased social costs of negative externalities on the environment and human health	• Crop Diversification • Increased fertilizer use efficiency • Balance application of all nutrients	• Output price reform • Removal of fertilizer subsidies
Increased pest buildup and pest-related yield losses	• Continuous rice monoculture • Increased asymmetry of planting schedules • Greater uniformity in varieties cultivated	• Increased pesticide use		• Improved varieties with host plant resistance • Appropriate varietal turnover • Adoption and use	• Removal of pesticide subsidies • Investments in farmer education

and Roger, 1995, for detailed evidence in Philippine rice ecosystems), the Integrated Pest Management (IPM) program has been vigorously pursued, particularly for rice. To make IPM more attractive, pesticides should never be subsidized since, as in the case of fertilizers, farmers would have no incentive to invest time in acquiring IPM skills. Removing all explicit and implicit subsidies on pesticides is essential to reduce pesticide use on farms.

With the progression toward global integration, the competitiveness of domestic cereal agriculture can only be maintained through dramatic reductions in the cost per unit of production. New technologies designed to significantly reduce the cost per unit of output produced, either through a shift in the yield frontier or through an increase in input efficiencies, would substantially enhance farm level profitability of cereal crop production systems. Increasing input use efficiency would also contribute significantly to the long-term sustainability of intensive food crop production and help arrest many of the problems described above.

REFERENCES

Abrol, I.P. (1987). Salinity and food production in the Indian sub-continent. In *Water and Water Policy in World Food Supplies*, ed. W.R. Jordan pp. 109-113, Texas A&M Univ. Press, College Station.

Ahmed, N. (1985). Fertilizer efficiency and crop yields in Pakistan. *Phosphorus in Agriculture* 89:17-32.

Cassman, K.G. and P.L. Pingali. (1993). Extrapolating trends from long-term experiments to farmers fields: the case of irrigated rice systems in Asia. In *Proceedings of the Working Conference on "Measuring Sustainability Using Long-Term Experiments."* Rothamsted Experimental Station, 28-30 April 1993, funded by the Agricultural Science Division, The Rockefeller Foundation.

Cassman, K.G., S.K. De Datta, D.C. Olk, J. Alcantara, M. Samson, J. Descalsota, and M. Dizon. (1994). Yield decline and nitrogen balance in long-term experiments on continuous, irrigated rice systems in the tropics. *Advances in Soil Science*, Spl. Issue.

Chambers, R. (1988). *Managing Canal Irrigation: Practical Analysis from South Asia*. Press Syndicate of the University of Cambridge, Cambridge, UK.

De Datta, S.K. (1981). *Principles and Practices of Rice Production*. Wiley-Interscience Publication, John Wiley & Sons, New York, USA.

Desai, G.M. (1988). *Development of Fertilizer Markets in South Asia: A Comparative Overview*. Mimeo. Washington, DC: International Food Policy Research Institute.

Desai, G.M. (1986). Fertilizer use in India: The next stage in policy. *Indian Journal of Agricultural Economics* 41: 248-270.

Desai, G.M. and V. Gandhi. (1989). Phosphorus for sustainable agriculture growth in Asia–an assessment of alternative sources and management. In *Proceedings of the Symposium on Phosphorus Requirements for Sustainable Agriculture in Asia and Pacific Region*. International Rice Research Institute, Los Baños, Laguna, Philippines, March 6-10.

Dogra, B. (1986). The Indian experience with large dams. In *The Social and Environmental Effects of Large Dams*. Goldsmith, E. and Hildyard, N. (eds.) Vol 2: 201-208, Wadebridge Ecological Center, London, UK.

Fujisaka, J.S., L. Harrington, and P. Hobbs. (1994). Rice-wheat in South Asia: systems and long-term priorities established through diagnostic research. *Agricultural Systems* 46: 169-187.

Harrington, L.W., M. Morris, P.R. Hobbs, V. Pal Singh, H.C. Sharma, R.P. Singh, M.K. Chaudhary, and S.D. Dhiman (eds). (1992). *Wheat and Rice in Karnal and Kurukshetra Districts, Haryana, India–Practices, Problems and an Agenda for Action*. Report from an Exploratory Survey, 22-29 March 1992. Haryana Agricultural University (HAU), Indian Council for Agricultural Research (ICAR), International Maize and Wheat Improvement Center (CIMMYT), International Rice Research Institute (IRRI).

Heisey, P.W., M. Smale, D. Byerlee, and E. Sonza. (1997). Wheat rusts and the costs of genetic diversity in the Punjab of Pakistan. *American Journal of Agricultural Economics* 79 (August): 726-737.

Heong, K.L., G.B. Aquino, and A.T. Barrion. (1992). Population dynamics of plant and leafhoppers and their natural enemies in rice ecosystems in the Philippines. *Crop Protection* 11: 371-379.

Hobbs, P. and Morris, M. (1996). *Meeting South Asia's Future Food Requirements from Rice-Wheat Cropping Systems: Priority Issues Facing Researchers in the Post-Green Revolution Era*. Natural Resource Group Working Paper 96-01. CIMMYT, Mexico.

Hobbs, P.R., L.W. Harrington, C. Adhikary, G.S. Giri, S.R. Upadhyay, and B. Adhikary. (1996). *Wheat and Rice in the Nepal Tarai: Farm Resources and Production Practices in Rupandehi District. No. AECO95A Annual report R95A 53*. Centro Internacional de Mejoramiento de Maiz y Trigo (CIMMYT), Mexico, DF (Mexico).

Kataki, P.K., P. Hobbs and B. Adhikary. (2000). The rice-wheat cropping system of South Asia: trends, constraints and productivity–a prologue. *Journal of Crop Production*, 3(2)#6: 1-26.

Lopes, A.S. (1980). Micronutrients in soils of the tropics as constraints to food production. In *Priorities for Alleviating Soil-Related Constraints to Food Production in the Tropics*. Jointly sponsored and published by the International Rice Research Institute, Los Baños, Laguna, Philippines and the New York State College of Agriculture and Life Sciences, Cornell University.

Moormann, F.R. and N. van Breeman. (1978). *Rice: Soil, Water Land*. International Rice Research Institute (IRRI), Los Baños, Laguna, Philippines.

Mustafa, U. (1991). *Economic Impact of Land Degradation (Salt Affected and Waterlogged Soils) on Rice Production in Pakistan's Punjab*. Ph.D. dissertation, College of Economics and Management, University of the Philippines at Los Baños (UPLB), College, Laguna, Philippines.

Pingali, P.L. and M.W. Rosegrant. (1998). *Intensive Food Systems in Asia: Can the Degradation Problems be Reversed?* Paper presented at the Pre-Conference Workshop 'Agricultural Intensification, Economic Development and the Environment' of the Annual Meeting of the American Agricultural Economics Association, held in Salt Lake City, Utah, July 31-August 1, 1998.

Pingali, P.L. and R.V. Gerpacio. (1997). *Towards reduced pesticide use for cereal crops in Asia. Economics Working Paper No. 97-04*. Centro Internacional de Mejoramiento de Maiz y Trigo (CIMMYT), Mexico, DF (Mexico).

Pingali, P.L., M. Hossain, and R.V. Gerpacio. (1997). *Asian Rice Bowls*: *The Return-ing Crisis*? IRRI/CAB International.

Pingali, P.L. and M.W. Rosegrant. (1998). *Intensive Food Systems in Asia*: *Can the Degradation Problems be Reversed*? Paper presented at the Pre-Conference Workshop 'Agricultural Intensification, Economic Development and the Environ-ment' of the Annual Meeting of the American Agricultural Economics Associa-tion, held in Salt Lake City, Utah, July 31-August 1, 1998.

Ponnamperuma, F.N. (1974). Micronutrient limitations in acid tropical rice soils. In *Soil Management in Tropical America*, eds. E. Bornemisza and A. Alvarado. Soil Science Department, North Carolina State University, Raleigh, pp. 330-347.

Postel, S. (1989). Water for agriculture: facing the limits. *Worldwatch Paper 93*, December.

Rogers, P., P. Lydon, D. Seckler, and G.T. Keith Pitman. (1994). *Water and Develop-ment in Bangladesh*: *A Retrospective on the Flood Action Plan*. Synthesis Report Prepared for the Bureau of Asia and the Near East of the U.S. Agency for International Development by the Irrigation Support Project for Asia and the Near East.

Rola, A.C. and Pingali, P.L. (1993). *Pesticides, Rice Productivity, and Farmers' Health–An Economic Assessment*. IRRI-World Resources Institute (WRI). Inter-national Rice Research Institute (IRRI), Los Baños, Laguna, Philippines.

Samad, M., D. Merrey, D. Vermillion, M. Fuchs-Casrsch, K. Mohtadullah, and R. Lenton. (1992). Irrigation management strategies for improving the performance of irrigated agriculture. *Outlook on Agriculture* 21(4): 279-286.

Stone, B. (1986). Chinese fertilizer application in the 1980s and 1990s: issues of growth, balances, allocation, efficiency and response. In *China's Economy Looks Toward the Year 2000*. Volume 1. Congress of the United States, Washington, D.C. pp. 453-493.

Tandon, H.L.S. (1987). *Phosphorus Research and Agricultural Production in India*. Fertilizer Development and Consumption Organization, New Delhi, India.

Synthesis of Systems Diagnosis: "Is the Sustainability of the Rice-Wheat Cropping System Threatened?" – An Epilogue

L. Harrington

SUMMARY. Rice-wheat systems in the Indo-Gangetic Plains may seem simple, but in reality they are enormously complex, with numerous productivity and sustainability problems, each embedded in a web of interactions. By unraveling and understanding these interactions, and harnessing their power, scientists working with the Rice-Wheat Consortium for the Indo-Gangetic Plains are fostering sustainable improvements in system productivity. Rice-wheat system problems include late sowing, low water and nutrient use efficiency, groundwater depletion (in some areas) or waterlogging and poor water control (in other areas), salinity and sodicity (in defined locales) and the build-up of problem weeds and possibly of pests and diseases. The low soil fertility scenario is due to the increasing micronutrient deficiencies, widening NPK ratios or imbalanced fertilizer use, and decreased use of organic sources of nutrients. At first glance, these problems seem random and unrelated but in fact, many of them are closely linked through system interactions. Late sowing, for example, reduces water and nutrient use efficiency, and restricts the time available within the rotation for break crops, thereby contributing to the build-up of problem weeds and to soil fertility decline. Reduced and zero tillage for rice and wheat improve

L. Harrington is Director, Natural Resource Group, CIMMYT, Lisboa 27, Apartado Postal 6-64, 06600, Mexico, D.F., Mexico (E-mail: l.harrington@cgiar.org).

[Haworth co-indexing entry note]: "Synthesis of Systems Diagnosis: 'Is the Sustainability of the Rice-Wheat Cropping System Threatened?'–An Epilogue." Harrington, L. Co-published simultaneously in *Journal of Crop Production* (Food Products Press, an imprint of The Haworth Press, Inc.) Vol. 3, No. 2(#6), 2001, pp. 119-132; and: *The Rice-Wheat Cropping System of South Asia: Trends, Constraints, Productivity and Policy* (ed: Palit K. Kataki) Food Products Press, an imprint of The Haworth Press, Inc., 2001, pp. 119-132. Single or multiple copies of this article are available for a fee from The Haworth Document Delivery Service [1-800-342-9678, 9:00 a.m. - 5:00 p.m. (EST). E-mail address: getinfo@haworthpressinc.com].

timeliness of sowing and, through these same system interactions, help farmers cope with multiple productivity and sustainability problems. *[Article copies available for a fee from The Haworth Document Delivery Service: 1-800-342-9678. E-mail address: <getinfo@haworthpressinc.com> Website: <http://www.HaworthPress.com> © 2001 by The Haworth Press, Inc. All rights reserved.]*

KEYWORDS. Cropping system, Indo-Gangetic Plains, productivity, rice, wheat

SYSTEMS OF APPARENT SIMPLICITY

To the uninformed, few cropping systems appear simpler than the rice-wheat systems of the Indo-Gangetic Plains of South Asia. Rice monoculture (or so goes the story) is followed by wheat monoculture, time and time again, year in and year out. No need to wait for the monsoon–irrigation takes care of water requirements. And no need to be concerned with pests, weeds and diseases–modern varieties and a suite of external inputs make these a thing of the past. In this favored, high potential environment, how can there be room for complex system interactions? It seems so straightforward!

But the simplicity of rice-wheat systems is only apparent. In reality, they are enormously complex, with numerous productivity and sustainability problems; each embedded in a web of interactions. Often these are not pure rice-wheat systems at all, but rather systems where rice and wheat are grown in short- or long-term rotations or associations with legumes, pulses, sugarcane, potatoes or other crops (Razzaque, Badaruddin and Meisner, 1995). Rotations vary across the landscape, and may change from one year to the next in the same location. Crop-livestock interactions add further complications–crop residue management, fodder requirements, farmyard manure use, and the effects these have on soil fertility and system productivity.

Over the past two decades, a number of diagnostic studies of rice-wheat systems have been conducted. Many of these have looked into system interactions, and the interplay among productivity and sustainability problems, their causes, and possible solutions, all in a systems perspective. On the principle that problems are easier to solve when their causes are understood, these studies have been used to help set rice-wheat system research priorities (Rice-Wheat Consortium for the Indo-Gangetic Plains, 1995).

In this paper, a summary is presented of some problems affecting the productivity and sustainability of rice-wheat systems. Causes of these problems are then described, and the interactions among problems and causes are traced. Implications for rice-wheat system sustainability are summarized, and a set of leverage points described. Leverage points are understood as interventions, technologies or practices, even if straightforward and simple, that harness system interactions and thereby contribute significantly to sustainable improvements in system productivity and natural resource quality.

RICE-WHEAT SYSTEM PROBLEMS

Productivity and sustainability problems affecting rice-wheat systems vary across the Indo-Gangetic Plains, some of them following a rough east-west gradient. In general, rainfall is higher, water control more difficult, and soils heavier towards the east. Within these systems, the rice crop typically is affected by pests and diseases, midseason moisture stress and (in direct seeded rice) weeds. The wheat crop generally is affected by late sowing, early season waterlogging in lower terraces with heavier soils, inadequate plant stands, competition from grassy weeds, late season heat and moisture stress, and problems of varietal replacement. Nutrient deficiencies are common for both crops, as are low water and fertilizer use efficiency (Fujisaka, Harrington and Hobbs, 1994).

Longer-term problems tend to include soil fertility decline (Bronson and Hobbs, 1996), a build-up of weeds, especially in wheat (Hobbs, Sayre and Ortíz-Monasterio, 1998; see also Malik, Gill and Hobbs, 1998) and possibly a build-up of pests and diseases. To these may be added water management problems. Groundwater depletion affects areas in the western Indo-Gangetic Plains where groundwater pumping is excessive (Gill, 1992; Harrington et al., 1992). In contrast, other areas experience poor drainage and waterlogging. Sometimes these are due to naturally poor drainage or poor irrigation system water control (especially in the eastern Indo-Gangetic Plains), and at other times to specific irrigation practices (Singh and Singh, 1995). Finally, some specific areas have become saline or sodic, though other such areas have been reclaimed (Bhargava, 1989).

Problems and Causes

At first glance, these productivity and sustainability problems seem random and unrelated. In fact, many of them are closely linked, as will be illustrated below. First, however, the notion of land types is discussed, because some problems and their corresponding causes vary across land types.

Land Types and the Toposequence

Rainfall, water control and soil texture tends to follow an east-west gradient across the Indo-Gangetic Plains. Water control and soil texture in specific locations, however, is also influenced by land type (i.e., lower, middle and upper terraces) within a toposequence. Though a land type may be known by different local names in different parts of the Indo-Gangetic Plains, its characteristics, uses and management are often similar (Harrington et al., 1993; see also Fujisaka 1990 for a purely rice example).

Lower terraces are characterized by heavier soils and relatively poor drainage and are more likely to be devoted to long duration traditional rice cultivars. The lowest of these may be flooded or wet all year. Middle terraces have somewhat lighter soils and fewer drainage problems, and typically are sown to modern varieties of rice and wheat, at times mixed with other crops. Upper terraces have the lightest soils of all, and tend to have greater agroecosystem species diversity. Here rice and wheat are sown, but also pigeon pea, sugarcane, and vegetables.

Understandably, at a given location, waterlogging is more likely to be a problem in lower terraces, and inadequate moisture in upper terraces. Because they are poorly drained, lower terraces are more likely to experience long turnaround time between rice harvest and wheat sowing. The distinction among lower, middle and upper terraces at times may be measured in mere centimeters of elevation–and one year's middle terrace may be another year's lower terrace if the monsoon is heavy. Land types categories, then, is relative and depend on fine distinctions in hydrology. Finally, it should be recognized that the proportion of lower, middle and upper terraces in a landscape will vary across the Plains–lower terraces predominate in the east and middle and upper terraces predominate in the west.

Timeliness and Tillage

It has long been recognized that timeliness of operations is critically important in rice-wheat systems (Byerlee et al., 1984). Time-related factors influencing system productivity include the transplanting of old rice seedlings, the late planting of wheat due to delayed rice harvest, and a prolonged turnaround period between crops. Given that wheat yields may fall by around 40 kg/ha per late-planted day, that water and nutrient use efficiency are reduced by late sowing, and that a third crop (e.g., mung) cannot be introduced when wheat is sown late, it is easy to understand why practices that foster timely sowing have been given considerable attention (Hobbs and Morris, 1996).

Short-duration rice varieties can lead to increases in total system productivity for late planted rice; zero tillage establishment of wheat may reduce turnaround time; while wheat varieties which are high yielding over a range of plant dates increase farmer flexibility in maintaining system productivity (Flinn and Khokhar, 1989). Among these, a short-duration rice variety is not always the most attractive option for farmers–for example, when long season, high value basmati rice varieties are grown (Sharif et al., 1989). Understandably, turnaround between rice harvest and wheat sowing is especially challenging in lower terraces, or in the eastern part of the Indo-Gangetic Plains. Poor drainage can delay wheat sowing even when rice harvest is timely.

New reduced and zero tillage and crop establishment practices–for rice as well as for wheat–promise to finally assure timely rice and wheat sowing. These practices have been shown in several locations to advance sowing dates, cut costs, raise yields, improve nutrient use efficiency, reduce weed germination (in some instances), save water, reduce fuel use (and thereby carbon emissions) and generate higher incomes for farm families. Practices have been developed for lower terraces (surface seeding, Chinese Hand Tractor seeder) as well as for middle and upper terraces (Chinese Hand Tractor seeder, Pantnagar drill). (See, for example, Meisner et al., 1999; Hobbs and Rajbhandari, 1998; Rice Wheat Consortium for the Indo-Gangetic Plains, 1995b. Much of the supporting information resides in very new data summaries, internal memoranda, and unpublished reports.)

Soil Fertility and Nutrient Use Efficiency

A major threat to the sustainability of rice-wheat system productivity comes from soil fertility loss. A number of studies have reported on the outcomes of long-term soil-fertility experiments (e.g., note especially Abrol et al., 2000). Most of these conclude that soil fertility loss does indeed take place in rice-rice-wheat and rice-wheat systems, that rice yields are affected more than wheat yields, and that high NPK levels do not entirely replace the need for organic amendments, e.g., farm yard manure.

Some studies conclude that a decline in soil organic matter is the most important process at work (Bronson and Hobbs, 1996; Nambiar, 1985). Other studies suggest that the deterioration in productivity is associated with deficiencies of secondary nutrients and micronutrients such as sulfur and zinc (Singh and Paroda, 1993). A recent Rice Wheat Consortium publication on long-term experiments (Abrol et al., 2000) notes that, "The yield of rice achievable with nutrient inputs was shown to drop over time in some experiments, suggesting that constraints other than macro-nutrients contribute to yield declines. Micro-nutrient deficiencies and the build-up of pests and pathogen pressures are possible additional constraints . . . " (p. xxi).

Whether the problem lies with soil organic matter and its nitrogen supplying capacity, or with secondary and micronutrients, many causes appear similar (see various rice-wheat system diagnosis reports, e.g., Ahmed et al., 1993; Fujisaka, Harrington and Hobbs, 1994; Harrington et al., 1993; Hobbs et al., 1992; Hobbs et al., 1991; and others).

- Reduced levels and frequency of farmyard manure applications to crop fields, attributable to changes in livestock herd size and composition (sizeable reduction in the number of draft animals, not fully compensated by a small increase in the number of dairy animals), combined with increased manure use for household fuel.
- Fewer crop residues retained in fields (whether incorporated or left as a surface mulch) attributable to increased use of crop residues for livestock fodder (linked to reductions in grazing/pasture area) and increased burning, especially of rice residues.
- Increased reliance on continuous rice-wheat rotations and a decline in the attractiveness of break crops, attributable to relatively

low yields and profits from break crops and/or market access problems (e.g., with sugarcane), these influenced by the policy environment.

- Relatively intensive nutrient mining in surface soil layers, attributable to restricted root growth associated with the plow pan created during puddle rice culture.
- Relatively low application levels of inorganic fertilizers, insufficient to replace nutrients extracted during crop production.
- Use of inorganic fertilizer sources that do not contain secondary or micronutrients.

Interactions between timeliness and tillage on the one hand, and soil fertility management on the other, are fairly obvious. Reduced and zero tillage practices reduce turnaround time between rice harvest and wheat sowing, thereby creating "space" in the system that may be used to establish a third, "break" crop. Given that soil fertility issues are concentrated in continuous rice-wheat systems, practices that make it easier for farmers to introduce break crops can improve soil fertility. The new tillage and establishment practices also facilitate better plant stands and timely sowing, improving nutrient use efficiency. Typically, these practices foster residue retention, with favorable impacts on soil organic carbon. Finally, and most obvious, the introduction of zero or reduced tillage practices normally requires substantial changes in how fertilizers are applied.

In contrast, it is unclear whether there is a link between land types and soil fertility issues. Lower terraces have heavier soils with more clay and silt. But upper terraces tend to have a greater incidence of break crops and may receive preferential treatment in farmyard manure applications (e.g., to high value vegetables). In one sense, the new rice and wheat tillage and establishment techniques may facilitate an extension of diverse rotations from upper to middle terraces.

Water Management

Threats from water management problems to the productivity and sustainability of rice-wheat systems in the Indo-Gangetic Plains may be even more important than those posed by soil fertility loss. Expansion of irrigated area and intensification of cropping patterns have increased the demand for water throughout South Asia. Concern over water availability has mounted especially fast in areas of northwestern

India and the Pakistan Punjab, where water tables have dropped at a rate of 0.2-1.0 m per year. This has already raised pumping costs, reduced irrigation frequency and placed limits on sown area (Hobbs and Morris, 1996). Poor quality tubewell water, salinity, sodicity, and poor drainage/waterlogging are problems in different parts of the Indo-Gangetic Plains.

Many water management problems have their roots in institutional arrangements, institutional performance, and policy decisions. These affect farm level water management practices, encouraging wasteful use. For example, when groundwater is perceived as common property ("if I don't pump it and use it, someone else will"), it is in the interest of an individual farmer to pump as much as possible–unless collective action arrangements are in place (e.g., restrictions on tubewell installation) (McCulloch et al., 1998). Subsidies on water and on electricity for tubewell pumping are also contributing to excessive water use. Some analysts recommend tradable water rights as a way to ensure efficient use of water resources, while allowing farmers to benefit from their traditional water rights (Rosegrant and Gazmuri, 1994).

The capacity to examine the feasibility of new practices, e.g., conjunctive use of surface water and groundwater, is hampered by lack of sufficient communication and coordination among institutions responsible for management of surface water, public tubewells, electricity supply, drainage, on-farm works, extension, and the reporting of agricultural data (Murray-Rust and Vander Velde, no date).

At the plot level, water management problems interact with tillage and establishment, and soil fertility management, in many ways (see Rice Wheat Consortium for the Indo-Gangetic Plains, 1995a). At times, these interactions are conditioned by land type:

- Poor field drainage can increase turnaround time between rice harvest and wheat sowing, delaying wheat sowing and reducing yields, especially in lower terraces. At times this can substantially restrict the area sown to wheat after rice.
- Poor leveling of fields and/or poor water control can lead to poor wheat stand establishment, reduced productivity, and reduced nutrient use efficiency–again in lower terraces.

- Puddled rice culture in light soils in upper terraces can call for an immense amount of water, feasible only with substantial subsidies.
- Excessive water use contributes to loss of applied fertilizer nutrients, polluting groundwater and reducing fertilizer use efficiency.
- Excessive water use, especially for rice, may result in an increase in groundwater levels, eventually leading to waterlogging and perhaps salinity.

Curiously, given the fact that groundwater levels are falling in some parts of the Indo-Gangetic Plains and rising in others, it may be most suitable for farmers to be parsimonious with water in some areas (to avoid waterlogging) and "sloppy" in others (to accelerate groundwater recharge).

Agroecosystem Health

Many observers express considerable concern about a possible decline in agroecosystem health (especially increased incidence of pests, diseases and weeds) for rice-wheat systems. Continuous rice-wheat, it is thought, fosters a build-up of pests, diseases and weeds, relative to more diverse rice-wheat systems featuring break crops. Yet, apart from grassy weeds in wheat after rice, the evidence is patchy at best. Regarding pests, apprehension is sometimes expressed about build-up of stemborers and nematodes in continuous rice-wheat systems. Savary et al. (1997) note that different crop management practices, including rotations, do affect the incidence of pests, diseases and weeds but do not conclude that an unremitting build-up occurs with continuous rice-wheat. Still, the fact that the evidence is patchy does not mean that such a build-up does *not* occur.

In many areas, e.g., Haryana, *Phalaris minor* is a very important weed in wheat after rice (Harrington et al., 1992). Left uncontrolled it can lead to very substantial yield reductions. The problem is exacerbated by its development of resistance to the herbicides normally used for its control (Malik, Gill and Hobbs, 1998). A change in crop rotations can help remedy this problem. But another, innovative way is the use of bed sowing systems developed in Mexico. With wheat sown on beds, the *Phalaris* (preferring wetter conditions) tends to become concentrated in the furrows. This allows control through mechanical cultivation (see Sayre and Moreno Ramos, 1997).

Issues of agroecosystem health, then, demonstrate at least one clear system interaction with tillage and rotations: Zero and reduced tillage, by fostering earlier sowing and more system "space," facilitates more diverse rotations, thereby indirectly helping deal with any build-up in pests, diseases and weeds that *may* be associated with continuous rice-wheat rotations.

LEVERAGE POINTS AND SUSTAINABLE IMPROVEMENTS IN PRODUCTIVITY

In previous sections, rice-wheat system productivity and sustainability problems were described, along with their causes, possible solutions, and relevant system interactions. Given this information, is it possible to conclude that current rice-wheat systems are not sustainable? Or that they can be made sustainable by introducing new system management practices?

Over the past decade, a lot of effort has gone into defining sustainability and developing ways to measure it. Some analysts have explored sustainable development in the broadest sense (Lélé, 1991), while others have attempted to develop unique sustainability indicators, whether focusing on factor productivity (Ehui and Spencer, 1990) or land quality (Smyth et al., 1993). The question remains whether any of these approaches can satisfactorily determine whether a particular agroecosystem system is or is not sustainable. Or even whether there are simple "yes" or "no" answers to such questions.

Factor productivity approaches, for example, aim to gauge whether changes in system productivity are higher or lower than can readily be explained by changes in input levels. A system is deemed sustainable if factor productivity trends are positive. But these can be positive (over a limited range of time) even in the presence of significant and irreversible resource degradation, e.g., when yield-increasing technical change is (temporarily) swift, or when labor productivity is dramatically increased, e.g., through mechanization (Sidhu and Byerlee, 1992). Moreover, land quality indicators are fairly blunt instruments, combining as they do into a single value the effects of a few critically important factors along with many others that may be quite immaterial.

A more pragmatic approach would clearly identify the most important threats to sustainable improvements in productivity, and facilitate efforts to understand and then reverse them. This has been the approach of the Rice Wheat Consortium. The RWC does not aim to answer the question, "Are rice-wheat systems sustainable?" but rather "What are the main threats to sustained improvements in rice-wheat system productivity and what can be done about these threats?"

In "doing something" about threats to sustainable productivity, the RWC has sought to identify leverage points. Leverage points are understood as interventions, technologies or practices, even if straightforward and simple, that harness system interactions and thereby contribute significantly to sustainable improvements in system productivity and natural resource quality.

The most important leverage points identified by RWC scientists are, not surprisingly, tillage and crop establishment practices, especially zero- and reduced-tillage practices for rice and wheat. These are important because of their central location in the midst of a web of system interactions, most of which were introduced earlier. In summary:

- Nutrient management and soil fertility decline–New tillage practices improve nutrient use efficiency (through improved timeliness of sowing). They also facilitate farmer experimentation with more diverse farming systems; if these include break crops, soil fertility can be improved.
- Agroecosystem health–New tillage practices (e.g., bed systems) can help overcome the problem of *Phalaris minor* in wheat, especially where the weeds have developed a resistance to the herbicides used for their control. Other tillage practices (e.g., zero till) results in fewer weeds germinating. Earlier concerns about build-up of stem borer in crop residues appear unfounded (Mushtaq Gill, personal information). Some reduced-tillage practices, however (e.g., surface seeding) may foster increased problems with root rots or leaf blight in wheat.
- Water management–Finally, new tillage and establishment practices (e.g., broadcasting of rice seedlings, laser leveling of fields) can result in dramatic reductions in water requirements, expanding the amount of water available for other uses.

Rice-wheat systems in the Indo-Gangetic Plains may seem to be simple, but in fact they are enormously complex, with numerous productivity and sustainability problems, each embedded in a web of interactions. By unraveling and understanding these interactions, and harnessing their power, RWC scientists are fostering sustainable improvements in system productivity–beginning with seemingly simple innovations in tillage and establishment.

REFERENCES

Abrol, I.P., K.F. Bronson, J.M. Duxbury, and R.K. Gupta. (2000). Long-term Soil Fertility Experiments in Rice-Wheat Cropping Systems. Rice-Wheat Consortium Paper Series 6, New Delhi, India: Rice-Wheat Consortium for the Indo-Gangetic Plains.

Ahmed, N.U., M.M. Alam, A.M. Bhuiyan, R. Islam, and M.U. Ghani. (1993). *Rice-Wheat Diagnostic Survey: In Greater Kushtia District of Bangladesh*, Bangladesh:Bangladesh Rice Research Institute, p. vi, 57 p.

Bhargava, G.P. (1989). *Salt Affected Soils of India: A Source Book*, New Delhi: Oxford & IBH Publishing Co. Pvt. Ltd. p. 261.

Bronson, K.F. and P.R. Hobbs. (1996). *The Role of Soil Management in Improving Yields in the Rice-Wheat Systems of South Asia*. Workshop on Long-Term Research on Soil, Water and Nutrient Management Within the CGIAR System, 26-27 July, 1996, Ohio State University, Columbus, Ohio.

Byerlee, D., A.D. Sheikh, M. Aslam, and P.R. Hobbs. (1984). *Wheat in the Rice-Based Farming System of the Punjab Implications for Research and Extension*. NARC/CIMMYT Reports Series No. 4. Islamabad: NARC/CIMMYT. p. 49.

Ehui, S.K. and D.S.C. Spencer. (1990). *Indices for Measuring the Sustainability and Economic Viability of Farming Systems*. RCMP Research Monograph No. 3. Nigeria: IITA:v, p. 28.

Flinn, J.C. and B.B. Khokhar. (1989). Temporal Determinants of the Productivity of Rice-Wheat Cropping Systems. *Agricultural Systems* 30:217-233.

Fujisaka, S. (1990). Rainfed Lowland Rice: Building Research on Farmer Practice and Technical Knowledge. *Agriculture Ecosystems and Environment* 33:57-74.

Fujisaka, S., L. Harrington, and P. Hobbs. (1994). Rice-Wheat in South Asia: Systems and Long-Term Priorities Established Through Diagnostic Research. *Agricultural Systems* 46:169-187.

Gill, K.S. (1992). *Research on Rice-Wheat Cropping Systems in the Punjab*. Punjab Agricultural University Monograph. Ludhiana, Punjab, India: Punjab Agricultural University.

Harrington, L., S. Fujisaka, P. Hobbs, C. Adhikary, G.S. Giri, and K. Cassaday. (1993). Rice-Wheat Cropping Systems in Rupandehi District of the Nepal Terai: Diagnostic Surveys of Farmers' Practices and Problems, and Needs for Further Research. Sustainability of the Rice-Wheat System in South Asia. Mexico, D.F.: CIMMYT/NARC/IRRI:vii, p. 33.

Harrington, L., M. Morris, P.R. Hobbs, V. Pal Singh, H.C. Sharma, R.P. Singh, M.K. Chaudhary, and S.D. Dhiman. (1992). *Wheat and Rice in Karnal and Kurukshetra Districts, Haryana, India Practices, Problems and an Agenda for Action*. HAU/ICAR/CIMMYT/IRRI:viii, p. 75.

Hobbs, P.R., G.P. Hettel, R.P. Singh, Y. Singh, L. Harrington, and S. Fujisaka. (1991). *Rice-Wheat Cropping Systems in the Tarai Areas of Nainital, Rampur, and Pilibhit Districts in Uttar Pradesh, India: Diagnostic Surveys of Farmers' Practices and Problems, and Needs for Further Research. Sustainability of the Rice-Wheat System in South Asia.* Mexico, D.F. CIMMYT:iv, p. 55.

Hobbs, P.R., G.P. Hettel, R.K. Singh, R.P. Singh, L. Harrington, V.P. Singh, and K.G. Pillai. (1992). *Rice-Wheat Cropping Systems in Faizabad District of Uttar Pradesh, India: Exploratory Surveys of Farmers' Practices and Problems and Needs for Further Research. Sustainability of the Rice-Wheat System in South Asia.* Mexico, D.F. CIMMYT:vi, p. 61.

Hobbs, P. and M. Morris. (1996). *Meetings South Asia's Future Food Requirements from Rice-Wheat Cropping Systems: Priority Issues Facing Researchers in the Post-Green Revolution Era.* NRG Paper 96-01. Mexico, D.F. CIMMYT:vii + 46.

Hobbs, P. and N.P. Rajbhandari (eds.). (1998). *Proceedings of the Rice-Wheat Research End-of-Project Workshop*, 1-3 October 1997, Hotel Shangri-La, Kathmandu, Nepal. CIMMYT., NARC, and the Rice Wheat Consortium for the Indo-Gangetic Plains. Mexico, D.F.: CIMMYT and NARC:105 p.

Hobbs, P., K.D. Sayre, and J.I. Ortíz-Monasterio. (1998). *Increasing Wheat Yields Sustainably Through Agronomic Means.* NRG Paper 98-01. Mexico, D.F. CIMMYT:v + p. 19.

Lélé, S.M. (1991). Sustainable Development: A Critical Review. *World Development* 19(6):607-621.

Malik, R.K., G. Gill, and P.R. Hobbs. (1998). *Herbicide Resistance in Phalaris minor–A Major Issue for Sustaining Wheat Productivity in Rice-Wheat Cropping Systems in the Indo-Gangetic Plains.* Rice-Wheat Consortium Paper Series 3. New Delhi, India: Rice-Wheat Consortium for the Indo-Gangetic Plains:pp. 36.

McCulloch, A.K., R. Meinzen-Dick, and P. Hazell. (1998). *Property Rights, Collective Action and Technologies for Natural Resources Management: A Conceptual Framework.* SP-PRCA Working Paper No. 1. IFPRI and CGIAR. Washington, D.C. CGIAR:i + 64 p.

Meisner, C.A., P. Hobbs, M. Badaruddin, M.A. Razzaque, G.S. Giri, and S. Justice. (1999). *Mechanical Revolution Among Small Landholders of South Asia: The Growing Use of Chinese Hand Tractors.* Dhaka: CIMMYT.

Murray-Rust, D.H. and E.J. Vander Velde. *Conjunctive Use of Canal and Groundwater in Punjab, Pakistan: Management and Policy Options.* 1 + 23, no date. Unpublished.

Nambiar, K.K.M. (1985). All India Coordinated Research Project on Long-Term Fertilizer Experiments and its Research Achievements. *Fertilizer News* 30(4):56-66.

Nambiar, K.K.M. (1995). Major Cropping Systems in India. In *Agricultural Sustainability, Economic, Environmental and Statistical Considerations.* V. Barnett, R. Payne and R. Steiner (eds.). John Wiley & Sons, Chichester.

Razzaque, M.A., M. Badaruddin, and C.A. Meisner. (1995). *Sustainability of Rice-Wheat Systems in Bangladesh: Proceedings of the Workshop*, Nashipur: BARI, 1995. vi + 113 pp.

Rice-Wheat Consortium for the Indo-Gangetic Plains. (1995). *Nutrient Management*

Research Issues for Improving the Productivity and Sustainability of Rice-Wheat Cropping Systems in the Indo-Gangetic Plains. New Delhi: RWC.

Rice-Wheat Consortium for the Indo-Gangetic Plains. (1995a). *Water Management Research Issues for Improving the Productivity and Sustainability of Rice-Wheat Cropping Systems in the Indo-Gangetic Plains.* New Delhi: RWC.

Rice-Wheat Consortium for the Indo-Gangetic Plains. (1995b). *Tillage and Crop Establishment Research Issues for Improving the Productivity and Sustainability of Rice-Wheat Cropping Systems in the Indo-Gangetic Plains.* New Delhi: RWC.

Rosegrant, M.W. and R. Gazmuri. (1994). *Reforming Water Allocation Policy Through Markets in Tradable Water Rights: Lessons From Chile, Mexico, and California.* EPTD Discussion Paper No. 6. Washington, D.C. IFPRI: p. 56.

Savary, S., R.K. Srivastava, H.M. Singh, and F.A. Elazegui. (1997). A Characterization of Rice Pests and Quantification of Yield Losses in the Rice-Wheat System of India. In: *Crop Protection,* Great Britain: ELSEVIER, pp. 387-398.

Sayre, K.D. and O.H. Moreno Ramos. (1997). *Applications of Raised-Bed Planting Systems to Wheat. WPSR No. 31.* Mexico, D.F. CIMMYT. WPSR No. 31:v + 31.

Sharif, M., J. Longmire, M. Shafique, and Z. Ahmad. (1989). *Adoption of Basmati-385: Implications for Time Conflicts in the Rice-Wheat Cropping System of Pakistan's Punjab.* PARC/CIMMYT Paper No. 89-1. Islamabad: PARC:x, p. 31.

Sidhu, D.S. and D. Byerlee. (1992). *Technical Change and Wheat Productivity in the Indian Punjab in the Post-Green Revolution Period.* CIMMYT Economics Working Paper 92-02. Mexico, D.F. CIMMYT:v, p. 25.

Singh, J. and J.P. Singh. (1995). Land Degradation and Economic Sustainability. *Ecological Economics* 15:77-86, 1995.

Singh, R.B. and R.S. Paroda. (1993). *Sustainability and Productivity of Rice-Wheat System in the Asia-Pacific Region: Research and Technology Development Needs. 1993. Paper Presented at the Regional Expert Consultation on the Sustainability of the Rice-Wheat Production in Different Agroecological Settings in Asia,* FAO, RAPA, Bangkok 6-9 July.

Smyth, A., J. Dumanski, G. Spendjian, M. Swift, and P.K. Thornton. (1993). *FESLM: An International Framework for Evaluating Sustainable Land Management.* World Soil Resources Report No. 73. FAO:vii, p. 76. ISSN 0532-0488.

Index

Actinomycetes, as augmentation of manure, 40-41
Agroecosystem health, 127-128
Agronomic benefits of legumes, 72-74
Animal nutrition, legumes in, 71-72

Bacteria, as augmentation of manure, 40-41
Bangladesh
area trend: rice, 4-6
area trend: wheat, 6-7
production trend: rice, 7-9
production trend: wheat, 9-11
yield trend: rice, 11-12
yield trend: wheat, 12-14
Berseem *(Trifolium alexandrinum)*, 73
Biotic constraints to legume culture, 80-83

Centro International de Mejorimiento de Maiz y Trigo (CIMMYT)
policy re-directions, 103-118
regional diagnostic and benchmark survey, 1-22
synthesis of systems diagnosis, 119-130
Climate, of Indo-Gangetic Plains, 53-65. *See also* Indo-Ganetic Plains
Climate prediction, 61-63
Cornell University
long-term yield trends in northwest India, 27-52
regional diagnostic and benchmark survey, 1-22
Crop yield. *See* Yield; Yield trends

Diseases, legume culture and, 81

Ecological consequences of intensification, 106-113
groundwater depletion, 108-109
increasing losses due to pests, 111-113
micronutrient deficiencies and toxicity, 110-111
reversing current degradation trends, 113-116
salinity and waterlogging, 107-108
soil-nutrient status, 109-110
soil physical characteristics, 111
Effective microorganisms, 40-41

Farmyard manure, crop yields and, 38-42
Fertilizer subsidies, 114
Forage legumes, 67-89. *See also* Legumes

Government price supports, 113-114
Grain legumes, 67-89. *See also* Legumes
Green manure, crop yields and, 38-42
Green manures, 67-89. *See also* Legumes
Green Revolution, legume culture and, 74-77
Groundwater depletion, 108-109

Human nutrition, legumes in, 69-71

India, 32
area trend: rice, 4-6
area trend: wheat, 6-7
production trend: rice, 7-9
production trend: wheat, 9-11
yield trend: rice, 11-12
yield trend: wheat, 12-14
Indo-Gangetic Plains, 2-3
agroclimatological characterization, 53-65
climate prediction, 61-63
climatic risk, 59-61
data source issues, 63-65
land and climatic patterns, 122
overview of climate, 54-58
world comparison of climate, 58-59
Insect damage, 111-113
legume culture and, 81
losses related to intensification, 111-113
Integrated Pest Management, 114-115
Intensification-induced degradation, 106-113
groundwater depletion, 108-109
increasing losses due to pests, 111-113
micronutrient deficiencies and toxicity, 110-111
reversing current trends, 113-116
salinity and waterlogging, 107-108
soil-nutrient status, 109-110
soil physical characteristics, 111
Irrigation
groundwater depletion and, 108-109
salinity and waterlogging issues, 107-108
Isoproturon™, 20

Labor shortages, 21-22
Lactic acid bacteria, as augmentation of manure, 40-41
Land preparation practices, in legume culture, 85-87
Land types and toposequence, 122

Legumes, 67-102
benefits from, 69-74
agronomic, 72-74
in animal nutrition, 71-72
in human nutrition, 69-71
constraints to productivity, 80-84
historical and cultural perspective, 68
options for rice-wheat culture, 79-80
grain legumes, 78-79
green manure and forage legumes, 79-80
production, availability, and accessibility, 74-78
productivity issues, 84-89
forage and green manure legume value, 88-89
land preparation, 85-87
seedling vigor, 87
soil fertility, 87-88
weeds, 88
Leverage points, 129-130

Manure
farmyard, 38-42
green, 38-42, 79-80. *See also* Legumes
Marketing issues, in legume culture, 84
Micronutrient deficiencies, 110-111
Microorganisms effective with manure, 40-41
Mining of nutrients, 19-20
Monitoring methods, 15-16

Nepal, 33
area trend: rice, 4-6
area trend: wheat, 6-7
production trend: rice, 7-9
production trend: wheat, 9-11
yield trend: rice, 11-12
yield trend: wheat, 12-14

Nitrogen, partial factor productivity and, 109
Nitrogen inputs, crop yields and, 34-37
Nutrient imbalance, 19-20
Nutrient imports, crop yields and, 34-38
Nutrition
    legumes in animal, 71-72
    legumes in human, 69-71

Organic inputs
    crop yields and, 38-41
    sustainability and, 41-42

Pakistan
    area trend: rice, 4-6
    area trend: wheat, 6-7
    production trend: rice, 7-9
    production trend: wheat, 9-11
    yield trend: rice, 11-12
    yield trend: wheat, 12-14
Partial factor productivity (PFP) analyses, 22-23
Pesticides, 112-113
Pest losses related to intensification, 111-113
Phalaris minor, 20,112,127. *See also* Weeds/weed control
Phosphorus (P)
    crop yields and, 37-38
    deficiencies of, 109-110
Phytates, 71
Policy issues, 21,103-118
    ecological consequences of intensification, 106-113
    groundwater depletion, 108-109
    increasing losses due to pests, 111-113
    micronutrient deficiencies and toxicity, 110-111
    salinity and waterlogging, 107-108
    soil-nutrient status, 109-110

    soil physical characteristics, 111
    historical background and trends, 104-106
    legume culture and, 67-89,84
    reversing current degradation trends, 113-116
    in water management, 126
Potassium (K)
    crop yields and, 37-38
    deficiencies of, 109-110
Power tillers, in legume culture, 85-87
Productivity. *See* Yield; Yield trends
Protein-energy malnutrition, legumes and, 70-71
Puddling, 111
Pulse crops. *See* Legumes

Rice-wheat cropping system
    apparent simplicity of, 120-121
    area trend: rice, 4-6
    area trend: wheat, 6-7
    characteristics of, 17-22
    legume options for, 78-80
    legumes and diversification, 67-102
        addressing poor productivity, 84-89
        benefits from legumes, 69-74
        conclusions, 89
        constraints to productivity, 80-84
        historical and cultural perspective, 68
        options for rice-wheat culture, 79-80
        production, availability, and accessibility, 74-78
    leverage points and sustainable improvements, 128-130
    policy redirections for, 103-118. *See also* Policy issues
    problems of, 16-21,121-128
        agroecosystem health, 127-128
        land types and toposequence, 122

soil fertility and nutrient use
efficiency, 123-124
timeliness and tillage, 123
water management, 124-127
production trend: rice, 9-11
production trend: wheat, 7-9
regional diagnostic and benchmark
surveys, 14-22
rotation practices in, 120,124-125
substitute crops in, 17-18
synthesis of systems diagnosis,
119-130
yield trend: rice, 11-12
yield trend: wheat, 12-14

Salinity, 107-108
Seedling vigor, of legumes, 87
Seed saving/storage, 21
Sesbania aculeata, as organic input,
38-42
Soil(s)
changes in nutrient status, 109-110
changes in physical characteristics,
111
fertility and nutrient use efficiency,
123-124
micronutrient deficiencies and
toxicity, 110-111
policy issues in reversing
degradation, 113-116
salinity and waterlogging, 107-108
Soil fertility, legume culture and,
87-88
Soil management, 18-19
Soil organic matter, 42-44
Soil properties, 30
organic content, 42-44
South Asia
area trend: rice, 4-6
area trend: wheat, 6-7
production trend:rice, 7-9
production trend: wheat, 9-11
yield trend: rice, 11-12
yield trend: wheat, 12-14

Straw, with green manure, 40-41
Sustainability, 14-15
defining, 128-129
leverage points and, 129-130
organic inputs and, 41-42
soil constraints on, 83

Timeliness issues, 123
Toposequence, 122
Total factor productivity (TFP)
analyses, 22-23
Toxicity, of soils, 110-111
Trifolium alexandrinum (berseem), 73

Varietal use, 21
Viral diseases, of legumes, 81

Waterlogging, 107-108
Water management, 20-21,124-127
Water subsidies, 114
Weeds/weed control, 20,112-113
legume culture and, 81,88

Yields
nutrient inputs and, 34-38
organic inputs and, 38-41
soil organic matter and, 42-44
sustainability of and organic inputs,
41-42
temporal trends in, 31-34
trends in northwest India, 44-47
Yield trends
long-term in northwest India, 27-52
rice, 11-12
wheat, 12-14

Zero tillage, 123, 125
Zinc deficiency, 110-111

T - #0574 - 101024 - C0 - 212/152/9 - PB - 9781560220855 - Gloss Lamination